PERSISTENCE AND CHANGE IN RURAL COMMUNITIES
A 50-YEAR FOLLOW-UP TO SIX CLASSIC STUDIES

PERSISTENCE AND CHANGE IN RURAL COMMUNITIES
A 50-YEAR FOLLOW-UP TO SIX CLASSIC STUDIES

Edited by

A.E. Luloff
Department of Agricultural Economics and Rural Sociology, The Pennsylvania State University, USA

and

R.S. Krannich
Department of Sociology, Utah State University, USA

CABI *Publishing*

CABI *Publishing* is a division of CAB *International*

CABI Publishing
CAB International
Wallingford
Oxon OX10 8DE
UK

CABI Publishing
10 E 40th Street
Suite 3203
New York, NY 10016
USA

Tel: +44 (0)1491 832111
Fax: +44 (0)1491 833508
E-mail: cabi@cabi.org
Web site: www.cabi-publishing.org

Tel: +1 212 481 7018
Fax: +1 212 686 7993
E-mail: cabi-nao@cabi.org

© CAB *International* 2002. All rights reserved. No part of this publication may be reproduced in any form or by any means, electronically, mechanically, by photocopying, recording or otherwise, without the prior permission of the copyright owners.

A catalogue record for this book is available from the British Library, London, UK.

Library of Congress Cataloging-in-Publication Data
Persistence and change in rural communities : a fifty year follow-up to six classic studies / edited by A.E. Luloff and R.S. Krannich.
    p. cm.
Includes bibliographical references and index.
   ISBN 0-85199-521-7
 1. United States--Rural conditions--Case studies. 2. Sociology, Rural--United States. 3.   Social change--United States. I. Luloff, A. E. II. Krannich, Richard S.
    HN59.2 .P466 2002
    307.72′0973--dc21
                          2002005742

ISBN 0 85199 521 7

Typeset in Optima by Columns Design Ltd, Reading, UK
Printed and bound in the UK by Biddles Ltd, Guildford and King's Lynn.

Contents

Contributors	vii
Acknowledgements	xi
1. Introduction	1
A.E. Luloff and R.S. Krannich	
2. Community Change and Community Theory	9
J.C. Bridger, A.E. Luloff and R.S. Krannich	
3. Sublette, Kansas: Persistence and Change in Haskell County	23
L. Bloomquist, D. Williams and J.C. Bridger	
4. Irwin, Iowa: Persistence and Change in Shelby County	45
E.O. Hoiberg	
5. Community Change in Harmony, Georgia, 1943–1993	71
G.P. Green	
6. Community Change and Persistence: Landaff, New Hampshire	95
F. Schmidt, E. Skinner, L.A. Ploch and R.S. Krannich	
7. Community and Social Well-being in Contemporary El Cerrito	117
R.S. Krannich and C. Eastman	
8. The Old Order Amish Community 50 Years Later	143
A.E. Luloff, J.C. Bridger and L.A. Ploch	
9. A 50-year Perspective on Persistence and Change: Lessons from the Rural Studies Communities	171
R.S. Krannich and A.E. Luloff	
References	179
Index	187

Contributors

Leonard E. Bloomquist is an Associate Professor of Sociology at Kansas State University. The focus of his research is on the sustainable development of rural communities in the Great Plains and other regions.

Leonard E. Bloomquist
Associate Professor and Head
Department of Sociology, Anthropology and Social Work
Kansas State University
204 Waters Hall
Manhattan, KS 66506–4003, USA
Phone: 785–532–4962
Fax: 785–532–6978
E-mail: bloomqui@ksu.edu

Jeffrey C. Bridger is a Senior Research Associate at the Institute for Policy Research and Evaluation at The Pennsylvania State University. His research focuses primarily on communities located in rural and rural–urban fringe areas. Specific interests include land use and land use change, sustainable community development, and conflicts over competing uses of natural resources. His recent research projects include a study of barriers to social connectedness in different community settings, an analysis of natural resource conflicts in forested regions, community responses to hazardous waste sitings, and barriers to sustainable community development in rural areas experiencing rapid social and economic change.

Jeffrey C. Bridger
141 Deepwood Drive

Pine Grove Mills, PA 16868, USA
Phone: 814–867–0468
E-mail: jcb8@psu.edu

Clyde Eastman is Associate Professor Emeritus of Development Sociology at New Mexico State University. His recent research focuses on agricultural labour, border issues, rural development, and small scale irrigated farms.

Clyde Eastman
2235 Ave de Mesilla
Las Cruces, NM 88005, USA
Phone: 505–523–0310
E-mail: ceastman@nmsu.edu

Gary Paul Green is Professor of Rural Sociology at the University of Wisconsin-Madison. Green received his PhD from the University of Missouri-Columbia and taught at the University of Georgia for 8 years. Over the past 8 years at the University of Wisconsin, his research and teaching have focused on community, economic and workforce development issues. He is currently studying the role of community-based organizations in providing job training. Green also has been involved in workforce and community development issues in international settings, such as Ukraine, South Korea and New Zealand. Starting in 2003, he will become the editor of *Rural Sociology*.

Gary Paul Green
Professor of Rural Sociology
Department of Rural Sociology
1450 Linden Drive
University of Wisconsin-Madison
Madison, WI 53706, USA
Phone: 608–262–9532
Fax: 608–262–6022
E-mail: gpgreen@facstaff.wisc.edu

Eric O. Hoiberg is Associate Dean for Academic Programs in the College of Agriculture and Professor of Sociology at Iowa State University. His research interests center on technology and technology transfer, natural resources, and the interface between rural community and the changing structure of agriculture.

Eric O. Hoiberg
Associate Dean for Academic Programs
College of Agriculture
Iowa State University
134 Curtiss
Ames, IA 50014, USA
Phone: 515–294–6614

Fax: 515-294-5334
E-mail: hoiberg@iastate.edu

Richard S. Krannich is Professor of Sociology and Forest Resources and Director of the Institute for Social Science Research on Natural Resources at Utah State University. His current research interests focus on the social changes that occur in rural communities affected by transitions from dependence on traditional natural resource commodity industries to amenity-based development patterns.

Richard S. Krannich
Professor of Sociology and Forest Resources
Institute for Social Science Research on Natural Resources
Utah State University
Logan, UT 84322-0730, USA
Phone: 435-797-1241
Fax: 435-797-1240
E-mail: rkranich@wpo.hass.usu.edu

A.E. Luloff is Professor of Rural Sociology and Senior Scientist, Institute for Policy Research and Evaluation at The Pennsylvania State University. His work focuses on the impacts of rapid social changes on the human and natural resource bases of the community. Changes in land cover and use, particularly at the rural–urban fringe, and the impact of rural development and environmental policy have been central features of his work. He has just completed a 4-year term as editor of *The Journal of the Community Development Society*.

A.E. Luloff
Professor of Rural Sociology and Senior Scientist, Institute for Policy Research and Evaluation
The Pennsylvania State University
114 Armsby Building
Penn State
University Park, PA 16802, USA
Phone: 814-863-8643
Fax: 814-865-3746
E-mail: ael3@psu.edu

Louis A. Ploch is Professor Emeritus of Rural Sociology at the University of Maine, Orono, where for many decades he carried the torch for the sociology of rural places, in both the classroom and meeting halls of northern New England. He served as mentor to many social scientists in Vermont, New Hampshire, and Maine. His work focused on rural community development and the impacts of migration on rural communities. He is a past recipient of the Distinguished Rural Sociologist Award from the Rural Sociological Society.

Louis A. Ploch
5 Sylvan Road

Orono, ME 04473, USA
Phone: 207–581–3153

Fred Schmidt is an Associate Professor in the Department of Community Development and Applied Economics at the University of Vermont. In 1978, he founded and currently co-directs the Center for Rural Studies, an entrepreneurial 'center of excellence' at Vermont's largest university.

Fred Schmidt
Center for Rural Studies
207 Morrill Hall – University of Vermont
Burlington, VT 05405, USA

Elizabeth Skinner holds an MS from the Department of Community Development and Applied Economics at the University of Vermont. A practicing applied economist, Elizabeth is particularly interested in sustained natural resource development. She currently directs product development at Shelburne Farms, Shelburne, Vermont.

Elizabeth Skinner
14 Spruce Court
Burlington, VT 05402, USA

Duane D. Williams is a Community Development Specialist with the University of Missouri Outreach and Extension Service. The focus of his work is in community leadership development, strategic planning, health and social services and community infrastructure. His current research interests include determining the characteristics of successful communities, applied demographics, and the implementation of community housing assessment protocols.

Duane D. Williams
Community Development Specialist
Nodaway County Outreach and Extension Center
Courthouse Annex, 305 N. Market Street
Maryville, MO 64468, USA
Phone: 660–582–8101
Fax: 660–562–2011
E-mail: WilliamsDD@missouri.edu

Acknowledgements

This book represents the combined efforts of a number of individuals. First, we extend our sincere appreciation to our colleagues and collaborators – Len Bloomquist, Jeff Bridger, Clyde Eastman, Gary Green, Eric Hoiberg, Lou Ploch and Fred Schmidt – who joined us in conceiving and carrying out the restudy of the six communities that were the focus of the original Rural Studies investigations. All of us wish to extend our sincere thanks to Lee Carpenter. We truly appreciate her tireless efforts at clarifying the occasionally obtuse written work set before her. On behalf of all of the collaborators in this restudy effort, we also extend our heartfelt thanks to the many people in each of the six study areas who gave unselfishly of their time and insights so that we might better understand the conditions and changes impacting their communities. Collectively, we also benefited from the hard work and insights of numerous undergraduate and graduate students at our respective institutions. They helped in a variety of ways – including data collection and analysis – and our sincere thanks go out to each of these individuals.

We also wish to acknowledge the debt that we and our discipline owe to the pathbreaking efforts of the authors of the original Rural Studies reports. The legacy of their analyses and insights motivated our efforts on this research project, and their contributions have provided guidance to several subsequent generations of rural community scholars.

In addition, we want to extend our thanks to Mr Tim Hardwick of CABI *Publishing*. Tim's gentle prodding and patience helped steer us towards completion of this project. He provided invaluable assistance. We also thank CABI *Publishing* for its work in producing this volume.

Finally, we acknowledge the important contributions that Kenneth P. Wilkinson made to this project. His work on rural communities served as a basis for much of what is presented in this text and contributed in major ways to our collective understanding of community and the forces that influence community change. Prior to his untimely death in 1993, Ken was an active participant in the restudy efforts that are the focus of this book. More importantly, he was our close friend and mentor, and his gentle demeanor and deep understanding of rural community life are sorely missed. We trust that we have honoured his efforts and contributed to his legacy.

Introduction

A.E. Luloff and Richard S. Krannich

This book addresses fundamental and still unresolved sociological questions: to what degree and in what ways do local communities persist as meaningful forms of social organization in contemporary rural America? This is an important disciplinary question, since the concept of community, the role of community in the social order, and the nature of community change have comprised central areas of focus in sociological theory and research since the founding of the discipline. The question is also one of substantial practical and policy relevance because localized social structures and processes linked to the concept of community have long been assumed to contribute in important ways to individual and collective well-being (see Wilkinson, 1991).

Like many areas of sociological enquiry, disciplinary attention to the social conditions and processes associated with the concept of community has ebbed and flowed over time, reflecting changing assumptions regarding the nature and importance of localized social structures in modern societies. Early sociologists devoted a great deal of attention to the effects of modernization, industrialization and urbanization on communal social relations (e.g. Lynd and Lynd, 1937; Wirth, 1938; Redfield, 1941; Tonnies, 1957). By the mid 20th century, some observers concluded that these broader forces of change had 'eclipsed' the local community as a meaningful unit of social organization (Stein, 1960), and that linkages to 'mass society' had largely undermined the relevance and capacity of localized institutions and organizations (Vidich and Bensman, 1958).

However, to paraphrase the words of Mark Twain, such sociological assertions regarding the demise of community appear to have been greatly exaggerated. Indeed, over the past decade or so there has been a dramatic

© CAB *International* 2002. *Persistence and Change in Rural Communities* (eds A.E. Luloff and R.S. Krannich)

resurgence of attention to, and interest in, community and community-based social processes, on a variety of fronts. Within the past several years the American Sociological Association has spawned a new, community-oriented journal, *City and Community*, and articles addressing community phenomena have appeared with increased frequency, both in disciplinary journals such as *Rural Sociology* and interdisciplinary journals such as *Society and Natural Resources*. For the past several years a 'community' focus has been incorporated into the programme themes of the Rural Sociological Society's annual meetings (e.g. 'The community effect in rural places', for that organization's 2002 meetings). In short, at the start of the 21st century it appears that sociology has rediscovered community as a core focus of enquiry.

Community-based social structures and processes have become a focus of increased interest and attention in non-academic circles as well. As Swanson (2001, p. 8) observed, recent years have witnessed a broad-based 'cultural revival' of interest in the 'significance of local society'. Community-based land-use planning efforts have become a focus of widespread grass-roots efforts to sustain local environments and natural resource conditions through the development of watershed councils and similar locality-oriented organizations (see Weber, 2000). Public officials and policy analysts have noted the central importance of locality-based civic engagement as a strategy for addressing a broad array of economic, environmental and social needs (see Kemmis, 1990). Community-based organizations and processes have become a centrepiece of federal resource management agencies' efforts to adopt more participatory decision-making strategies (Kruger and Shannon, 2000; Overdest, 2000), and of resource management strategies, such as those designed to promote 'community-based forestry'.

Clearly, the kinds of social structures and processes that comprise the local community remain highly relevant from a disciplinary perspective and in practical application. To fully understand and assess these phenomena requires an examination that takes into account the broad, and at times dramatic, patterns of social and economic change that have buffeted communities over the past century. The chapters that follow address the extent to which local social processes continue to exert their influence through an analysis of community persistence and change in six rural locales situated in several regions of the USA. Our work capitalizes on the fact that these same six locales were the focus of in-depth sociological studies conducted during the early 1940s. These six classic case studies provide a benchmark against which our analyses, conducted a half-century later, can be compared and contrasted to evaluate the ways in which the local community has evolved and the degree to which it remains a relevant form of social organization in contemporary rural America.

The Original Rural Life Study Series

In 1940 the United States Department of Agriculture (USDA) coordinated sociological case studies of six different rural communities: El Cerrito, New Mexico; Sublette, Kansas; Landaff, New Hampshire; the Old Order Amish of Lancaster County, Pennsylvania; Irwin, Iowa; and Harmony, Georgia (Fig. 1.1). These studies represent seminal works in the study of rural and community life in the USA at a critical point in history immediately following the Great Depression, when long-established agrarian social and economic systems were rapidly giving way to the forces of industrialization and urbanization. Many small-town observers realized that fundamental changes in rural communities had occurred as a result of the devastating effects of the Great Depression. While some communities experienced massive outmigration and economic decline, others enjoyed a period of moderate growth. In almost no case were agriculturally based communities spared substantial losses.

Unlike the more well-known American community studies of the era that examined specific aspects of social and economic change,[1] the Rural Life Study series was designed to provide a holistic picture of community change in six ideal typical communities. As Carl Taylor explained in the introduction to the series, entitled *The Culture of a Contemporary Rural Community*, the purpose of this investigation was:

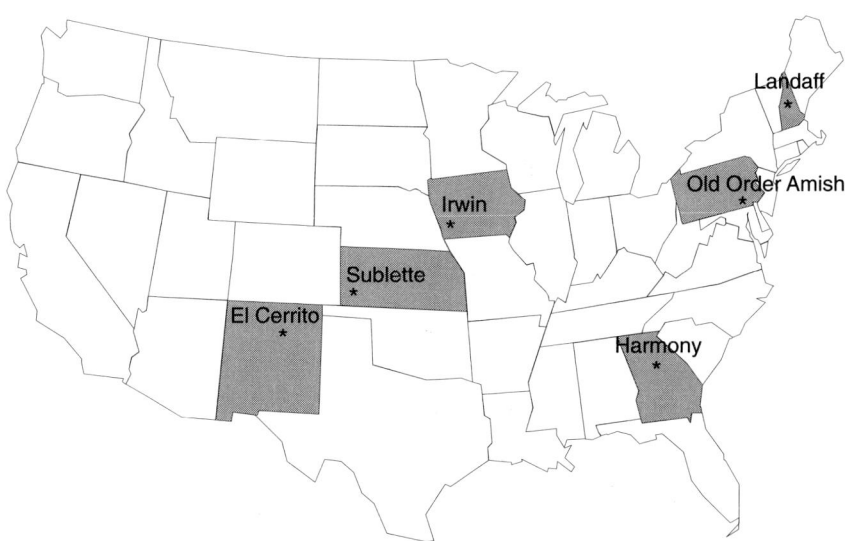

Fig. 1.1. Rural Life Study series research sites, 1941–1943.

> ... to investigate the cultural, community, and social psychological factors in land use and rural life, with special reference to those factors which either facilitate or offer resistance to change, contribute to adjustments and maladjustments, and to stability and instability in the individual and community life.
>
> (Taylor *et al.*, 1940, p. 1)

Each of the six communities was selected on the basis of a perceived degree of internal stability, ranging from Pennsylvania's Old Order Amish community at the highest (most stable) end to Sublette, Kansas, a dust-bowl community, at the other (the most unstable). The ethnographic research methods employed in this series of studies mirrored those used in most other community case studies conducted by sociologists during the early and mid-1900s, with a great deal of reliance placed on participant observation. The authors of the case-study reports spent a considerable time in the communities, living and interacting with residents. They also made use of surveys as well as secondary data such as census records and other public documents.

A series of publications on each of the six communities identified in this effort was published by USDA between 1940 and 1943. These studies were extraordinarily important because they stimulated the development of interest in the sociological study of rural communities. As reported by Ploch (1989, p. 1): 'Each study offers keen insights into the causes and effects of social changes that occurred over time.' Subsequent to their publication these community studies were frequently drawn upon by other community scholars to illustrate patterns of change affecting American communities (e.g. Warren, 1963, 1978) during the 20th century.

The Need for Comparative and Longitudinal Analysis

Despite the breadth and depth of material contained in the original case studies, their comparative potential was never fully realized, because, as Taylor noted (1945, p. 439):

> It was discovered by experience that the time allotted was not sufficient to accomplish the complete analysis ... the six community studies were, however, published with the recognition that they had only partially fulfilled the design and scope of analysis laid down in the manual of procedure.

In the years since the original effort, a few scholars have attempted to draw out comparative analyses based on the original cases studies. For example, Locke (1945) used data from the six communities to assess the condition of the rural family. After classifying the communities on such factors as family unity, family stability and familism, he concluded that the Old Order Amish and the residents of the geographically and culturally isolated community of El Cerrito ranked highest along these dimensions, with Irwin and Sublette occupying a middle-range position, and Harmony and Landaff ranking the lowest. Landaff, with its emerging value emphasis on individualism and the proportionately greater impact of urban culture, was observed as the most dissimilar along the dimensions identified.

In addition, Gross (1946) used observations from four of the communities (the Old Order Amish, El Cerrito, Irwin and Sublette) as the basis for his dissertation research and later published two articles from the same database (1948a,b). He relied on a comparative analysis to challenge the popular academic argument that communities could be classified meaningfully according to a rather crude rural–urban dichotomy. Based upon comparisons of the communities on such factors as cultural isolation, family systems, intracommunity interaction systems and religious systems, Gross argued that the uncritical use of the rural–urban dichotomy might lead observers 'to minimize the differentials within rural life and implies that because one is dealing with agriculture, a whole series of factors necessarily and ubiquitously follow' (Gross, 1948a, p. 256). To demonstrate how it was possible to reach such a deterministic conclusion, Gross pointed out that in crude terms, El Cerrito and the Old Order Amish appeared to form one kind of community while Sublette and Irwin formed another. Closer inspection revealed that cultural isolation helped to explain why the Old Order Amish and El Cerrito differed from Sublette and Irwin in the degree of familism, the importance of religion and educational systems, and the prevalence of secondary ties.

The importance of Gross's work is that it demonstrates how a comparative approach to community studies can lay bare important differences that are missed by other methodological techniques. While those who write textbooks on community have emphasized this point (cf. Warren, 1963, who focuses on El Cerrito as one of four key cases used in elaborating his thesis; Poplin, 1979; Sanders, 1950), there remains very little community research which systematically compares two or more communities along a series of sociological dimensions.[2]

The potential of the original studies to provide historical benchmarks for subsequent analyses of these same communities has also not been fully realized. Several individuals have attempted partial restudies. Loomis (1958, 1959) and Nostrand (1982) both reported on changes in El Cerrito, New Mexico; Mays (1968) conducted a restudy of Sublette, Kansas; and Ploch (1989) restudied Landaff, New Hampshire. On the other hand, a series of contemporary research articles on the Old Order Amish of Lancaster, Pennsylvania, by Buck (1978), Ericksen *et al.* (1980), Ericksen and Klein (1981), and Olshan (1981) failed to use the earlier material reported by Kollmorgen (1942).

While there was a general absence of restudy effort on the original six communities, the broader community studies literature does provide some important insights regarding the importance of longitudinal efforts. James West first studied Plainville (1945) in an attempt to understand the influence of mass society on a small agricultural community. Fifteen years later,

Gallaher (1961) restudied Plainville, using a similar methodology and format, with the explicit purpose of determining the forces of social and cultural changes and their impact on the community. Similarly, the processes and influences of industrialization on the small community were elaborated in a series of studies on Middletown (Lynd and Lynd, 1929, 1937), while Caplow et al. (1982, 1983) provided much data on familial and religious changes and continuities in the same community 50 years after the Lynds' work. Summers and Clemente (1976) provide additional insights into the effects of rural industrialization through an analysis of longitudinal data from a rural Illinois region. Other studies which provide a wealth of information on comparative change, especially with respect to community structure and dynamics, include the work of Warner and associates (1949) on Jonesville and Hollingshead (1950) on Elmstown (both of which were the same New England community), Ploch (1951) on Howard, Pennsylvania, and the work of Brunner and associates on 140 village-centred communities (Brunner et al., 1927; Brunner and Kolb, 1933; Brunner and Lorge, 1937).

Until now, no other researchers have attempted to fulfil the charter of the original study by conducting a systematic, comparative restudy of all six of the original Rural Studies series communities. Despite the fact that the thematic foci pursued in the original studies – community and family stability, agricultural and natural resource dependency, land use and environmental issues (drought, erosion, water quality) – remained relevant, little effort was expended on revisiting these study areas. This book addresses this shortcoming by examining the responses of these six small rural communities to the rapid and substantial changes of the last half of the 20th century, and evaluating their status at the end of the century.

The Case Study Method in Community Research

Community studies have long been a central feature of American social science. The dominant, early sociological trend was towards the ethnographic study of site-specific localities of relatively small population size. Such case studies were not only rich in detail, but contributed many significant theoretical and methodological advances to the discipline. However, with the advent of high-speed computers, large archives of relatively accessible secondary data, and advances in methodology and social statistics, attention turned to the quantitative, comparative study of urban places. At the same time, in-depth case studies became far less common and, although this shift allowed researchers to study macro-level trends easily, the depth and richness provided by detailed case studies was lost. Perhaps more importantly, the reliance on secondary data sources contributed to the false belief that the small and rural community was nothing more than a microcosm of its larger, more urban counterpart.

This conclusion was also fuelled by proponents of the mass society thesis who argued convincingly that the isolated, self-sufficient community, where people lived from cradle to grave, was gone forever. Advances in communication and transportation linkages were rapidly developing. Spreading from central city to central city, these networks provided a means for integrating much of the rural hinterland which had been bypassed in the past. Rapid printing was also important, for it allowed the dissemination of printed material to large numbers of people at relatively low costs. At the same time, such printing facilitated the dilution of the intellectual and artistic levels of society so as to appeal to a broader audience (cf. Bauer and Bauer, 1960). This dilution is still lamented, especially by critics of television, the medium perhaps most responsible for the levelling of attitudes, customs, mores and symbols in society. Through these processes people were being brought closer together and greater interdependencies formed, yet paradoxically mass society also implied a greater estrangement among individuals. Traditional networks, often rooted in family and local community ties, and characterized by parochial faiths, were lost, and replaced by a more secular, non-local societal order (Bell, 1956). Unfortunately, many of the consequences attributed to the massification of society were more firmly rooted in assertion and assumption than in attention to empirical details.

The use of computers, advanced methodological and statistical techniques, coupled to questionable theoretical assumptions about social change, led a deluge of rather muddled community research. Many of these studies examined very limited aspects of community life and lacked a sensitivity to the problems of real people. Not surprisingly, much of our best knowledge of the ramifications of major social change on the local community still comes from the earlier community case-study ethnographies.[3]

It would be unfair to place full blame for the abandonment of in-depth community case studies on today's well-trained, empirical researchers. Several of the authors of these same ethnographies, especially the Lynds and Warner, subscribed to the philosophy that by examining a typical community, enough detail could be assembled to generalize information for other communities with similar characteristics. Using parallel arguments, many of today's macro-modellers have suggested that the size of the sample enables their generalizations for other communities. Both perspectives have contributed to the false belief that the small and rural community is nothing more than a mere microcosm of its larger, more urban counterpart (cf. Luloff, 1990; Luloff and Wilkinson, 1990), and that little, if any, attention needed to be directed to rural community studies.

The Rural Life Study series recently celebrated its golden anniversary. Several generations of community sociologists were trained during and soon after these efforts, and many have either retired or are nearing retirement. These social scientists provided the fabric for many of today's studies in the form of concepts drawn from their efforts, research designs and sites for study. Given the need for more detailed understanding of local social organization and response to social change, the greatest acknowledgment of our debt to these researchers would be the requirement of field research for all community graduate students. Knowledge of empirical reality can only come from existential experience.

Organization of the Volume

In the remaining chapters of this book our goal is to draw upon insights from the re-examination of the six Rural Community Study areas to document some of the diverse patterns of social organization and change that have characterized American rural communities over the 50 years following the original study effort. In Chapter 2 we present an overview of major theoretical perspectives that provide a foundation for addressing these issues, focusing on the seminal works of Roland Warren (1963, 1978) and Kenneth Wilkinson (1991). Chapters 3–8, which present individual case-study reports for each of the six communities, provide an examination of contemporary community conditions as juxtaposed against the conditions that were documented in the early 1940s and the changes that have ensued since that time. Chapter 9 draws upon the findings from all six of the case-study chapters in developing a comparative assessment of the patterns of change that have occurred, the effects of change on community capacity and localized social interaction processes, and the implications of such changes for rural community well-being and rural development policy.

Notes

[1] These include Middletown (Lynd and Lynd, 1929), Yankee City (Warner, 1963), studies of the Deep South (Davis *et al.*, 1944) and Plainville (West, 1945).

[2] Some early exceptions include the work of Carter (1946, 1947a,b), Hay *et al.* (1949), Miner (1949), DuWors (1952) and Vogt and O'Dea (1953). However, in general, the current community literature is not marked by efforts at systematic, comparative study.

[3] The work of Caplow, Bahr, Chadwick and associates (Caplow *et al.*, 1982, 1983) represents an exception to this pattern. Their effort in Middletown from spring 1976 to autumn 1978 was largely a replication of earlier work. In their study, more attention was given to the use of surveys than in the Lynds' work, but they engaged in participant observation and made use of much secondary data in the form of records and documents.

Community Change and Community Theory

Jeffrey C. Bridger, A.E. Luloff and Richard S. Krannich

Introduction

For most of the 20th century, community theory was dominated by approaches that emphasized the stable and predictable aspects of local life. Influenced heavily by the systems theorizing of Talcott Parsons, many scholars defined community as a social system characterized by enduring patterns of structured interaction between two or more units.[1] As a system of interacting member units, the community was clearly distinguished from the surrounding environment. According to Roland Warren (1978, p. 145), 'the system endures as long as these units remain in a systematic relationship, as differentiated from the relation of the units to other units in the environment.' To ensure its identity as a unique system, each community performed a set of conscious and unconscious processes called boundary maintenance, which ranged from such rational procedures as the designation of spatial boundaries to local norms and colloquialisms.

Boundary maintenance alone, however, was not enough to ensure the systemic relationship among the different parts of the community. Internal and external changes constantly threatened to disrupt well-established patterns of life. Somehow, though, most communities managed to function relatively smoothly. According to systems theory, the concept of equilibrium made this possible. Except in the rarest of circumstances, such as war or famine, a community was always in a state of equilibrium. Communities naturally sought to 'react to a change in such a way as to minimize that change's impact on the relation of the units in the system' (Warren, 1978, p. 147).

By the 1950s and 1960s, it was apparent that the economic, social, cultural and technological changes unleashed by the Second World War were radically reshaping life at the local level. Suburban sprawl was obliterating community boundaries. The rise of television and other forms of mass media were flattening regional cultural distinctions, and advances in transportation and communication enabled people to participate in multiple communities.

Given the scale and pace of change, it is not surprising that many scholars turned their attention to understanding how community life was being altered. At first, attention focused on how change affected the community as a social system. Later, many students of community concluded that the systems model was no longer relevant to conditions in American communities and abandoned it in favour of theoretical approaches that could more easily accommodate change.

In this chapter, we describe the most influential theory of community change that emerged during this period, namely, Roland Warren's 'Great Change' thesis (Warren, 1963, 1972, 1978). Following this, we assess the accuracy of his thesis and its implications for community theory. We conclude with a discussion of contemporary approaches to community and social change.

The Great Change

Over the past six decades, the trends identified in the original Rural Life Study series have intensified, resulting in what Roland Warren (1963, 1972, 1978) called the 'Great Change' in community life. Like Tonnies's (1957) distinction between *gemeinschaft* and *gessellschaft* and Durkheim's (1947) discussion of mechanical and organic solidarity, the Great Change thesis can be seen as an attempt to impose a sense of order and coherence on a messy set of historical processes. Warren argued that as communities became increasingly reliant on extra-local institutions and sources of income, a sharp decline in autonomy and local solidarity occurred. One of the major trends affecting local life over the course of the 20th century has been the progressive intrusion of extra-local governmental and corporate systems into community affairs. In the course of this process, the locus of decision-making shifted to people and places far removed from the local scene. While decisions, policies and programmes that affect communities must still conform, in some respects, to local customs, norms and demands, many are now formulated outside the community and are 'guided more by their relations to extracommunity systems than by their relations to other parts of the local community' (Warren, 1978, p. 52).

While these external controls were being imposed, communities were becoming more internally differentiated as advances in travel and communication technology enabled people to develop relationships based on factors other than residential proximity. By the end of the First World War, Henry Ford had perfected the assembly line; soon, automobiles were affordable to the masses. With the creation of the interstate highway system and the growth of commercial air travel in the decades following the Second World War, Americans became the most mobile people in history. Geographic limits on interaction were lowered even further when the telephone service became widely available. Today, a whole new generation of technologies allows people to communicate with one another across vast distances.

Taken together, these developments have greatly expanded our interactional opportunities. We can now shop in one community, work in another, go to the doctor in a third, and live in a fourth. Nor do our friendships have to be tied to a particular place; communication technologies allow us to maintain close relationships with people who may live halfway around the world. As Wilkinson (1990, p. 153) put it, 'The local community is out; space-free networks are in', and, while it is arguable that technological advances have liberated people from the oppressive chains of small-town life described by Sherwood Anderson (1919) in his novel *Winesburg, Ohio* (Anderson, 1960), freedom has come with a high price tag – the relationships that held communities together have become fragile and transitory. As these localized or 'horizontal' linkages (Warren, 1978) have weakened, local solidarity has eroded, leaving communities bereft of the resources necessary to counter the macrostructural shifts described above.

We now live in communities that are very different from the self-sufficient towns surrounded by agricultural hinterlands that figure so prominently in our national mythology. According to the Great Change thesis, the community as a unit of social organization has become less important. Local autonomy has given way to dependence on external organizations and authorities, and residents and local organizations are oriented to happenings in the larger society and less interested and involved in community affairs.

The Great Change is probably best thought of as a shorthand description for a transformation of community life that has unfolded along several dimensions. As identified by Warren, these include: (i) division of labour; (ii) differentiation of interests and association; (iii) increasing systemic relationships to the larger society; (iv) bureaucratization and impersonalization; (v) transfer of functions to profit enterprise and government; (vi) urbanization and suburbanization; and (vii) changing values. Although not all communities have experienced these changes to the same extent or at the same pace, collectively they have caused a fundamental shift in patterns of local life.

In 1900, most Americans lived and worked on farms, and this common experience was a powerful source of community cohesion; farming provided a shared universe of discourse, a common relationship to the land, and a broadly shared set of interests arising out of living and working together. Today, less than 2% of the population are directly involved in agricultural production, and while this shift has resulted in highly complex division of labour and greater functional interdependence – we now rely on a host of people and organizations to provide the goods and services that were once produced at home, it has worked to reduce community solidarity. Lacking the occupational bond that once held communities together, residents are left without the diverse ties upon which local solidarity depends.

The second aspect of the Great Change, differentiation of interests and association, is closely related to the increasingly complex division of labour and the mobility associated with advances in transportation and communication. In pre-industrial America, interaction was limited to a small geographic area. People developed relationships with their neighbours because they had no other choices. In contemporary communities, the locality is no longer the unifying feature of life. Instead, people develop relationships primarily on the basis of shared interests, and this has further fragmented communities and reduced local solidarity:

> As association with neighbors declines, individuals often find themselves strangers in their own localities, knowing few if any of the neighbors. They find themselves interacting with people in *categorical* relationships (lawyer–client, salesperson–customer, home owner–plumber) rather than in *personal* relationships.
>
> (Warren, 1978, pp. 59–60)

The growing anonymity of community life has been blamed for a host of social ills, ranging from widespread alienation and anomie, to the difficulty of enforcing norms and controlling deviance through traditional mechanisms such as gossip, to the declining ability of community institutions to meet the needs of residents (Warren, 1978).

The division of labour and differentiation of interests have combined to create a new set of relationships between various parts of the community and larger systems. In the past, most businesses and voluntary organizations were locally based. Today there are very few strictly local entities. The branch plant of a large corporation, for instance, belongs partly in the community and partly within the larger organization of which it is a member. Moreover, as organizations have grown larger, they have centralized decision-making power at the state or national level. In the course of this process, the various parts of the community have become increasingly subservient to these larger systems.

As such extra-local or 'vertical' ties have increased in strength and number, the community's horizontal pattern – that set of relationships among parts of the local community – has suffered. Both individuals and local organizations are placed in a position where their behaviour must conform to the expectations of the parent organization, even when these expectations conflict with the needs of the community. The breakdown of the community's horizontal pattern, coupled to the transfer of decision-making power to extra-local entities, has also eroded community autonomy. This is most obvious when organizational goals transcend local boundaries, as in the case of a national university that sees its role in terms of its contribution to higher education in general rather than meeting the educational needs of the community in which it is located (Warren, 1978, p. 65). Even where organizational goals are relevant to the local community, as in the case of a chain store or medical facility that is affiliated with a larger hospital system, procedures and decisions become so standardized that they cannot be meaningfully tailored to the needs and wants of local citizens.

The divergence between organizational goals and community needs points to the fourth aspect of the Great Change – bureaucratization and impersonalization. Bureaucracies, with their impersonal rules and strict criteria for decision making, are essential to the efficient operation of the large-scale organizations that dominate social and economic life. At the same time, they have transformed relationships in ways that have profoundly affected community life. On the one hand, the growth of bureaucracy can help to ensure a level of equality that was absent when decisions were made on a more personal basis. On the other hand, impersonality sometimes comes at the expense of hurting a friend or neighbour, or results in decisions that do not accommodate special needs or circumstances. Impersonality in short can undermine relationships and patterns of social interaction that in the past made a community more than a mere aggregation of individuals.

Although the fifth aspect of the Great Change, the transfer of functions to profit enterprise and government, is related to all of the changes described above, it grows most directly out of the division of labour. By definition, the division of labour involves specialization. As people specialize, they rely on other people to perform tasks they once performed themselves. Individuals now either pay businesses directly for goods and services, or they pay indirectly to the government through taxes. What is important about this shift is that it signalled a movement away from reliance on such primary groups as family, friends and neighbours, toward business and government. As more activities were transferred to formal organizations, the building blocks of community – families, neighbourhoods and friendship groups – lost some of their vitality and reason for being (Warren, 1978, p. 74).

The Great Change was accompanied by a profound shift in settlement and land-use patterns. Between the first census in 1790 and the most recent count conducted in 2000, America went from a country where only 5% of the population lived in places of population 2500 or more, to one where over 80% of the population resides in cities and their surrounding suburbs. In fact, urbanization, and more recently, suburbanization, have been such ubiquitous processes that they are almost synonymous with the

Great Change. Consider, for instance, Louis Wirth's (1938) famous essay on urbanism as a way of life. In his attempt to develop a definition of the city, Wirth singled out three characteristics – large numbers of people, a dense settlement pattern and a heterogeneous population. Two decades later, Richard Dewey (1960) concluded that these aspects of urbanism produced a set of consequences remarkably similar to the five aspects of the Great Change discussed thus far. Urban life was characterized by specialization, anonymity, heterogeneity and impersonality. It was the antithesis of community – or at least that version of community rooted in the mythology of small-town America.

America's first suburb, Llewellyn Park, New Jersey, was completed in 1858. Since then, suburban life has been held up as an alternative to the crowding, crime, noise, filth and alienation of city life. At first this movement was limited to the wealthy, who settled in elite residential enclaves like Llewellyn Park that were connected to the city by commuter rail lines. With the growth of the interstate highway system and easy access to mortgage financing after the Second World War, suburban life was suddenly within reach of the masses. As a result, cities began to decant their populations so swiftly that many became dwarfed by the suburban ring that surrounds them.

Unfortunately, suburbs have in many ways reproduced the ills they were meant to overcome. First, unlike cities and towns, suburbs did not develop organically – they were planned carefully as residential settings with clearly segregated uses. The development pattern that dominated towns up until 1945, one characterized by 'mixed use, mixed income, apartments and offices over the stores, moderate density, scaled to pedestrians, vehicles permitted but not allowed to dominate, buildings detailed with care, and built to last' (Kunstler, 1996, p. 37) is largely absent from suburbia. Suburbs lack most of the features essential to the development of a holistic community (Kunstler, 1996). Further, precisely because modern suburbs were designed to accommodate the automobile, they typically lack the public places and pedestrian-friendly spaces that fostered community interaction. Finally, as Kunstler points out, suburbs represent the idea, borne of our experience with an immense frontier, that our problems can be solved by moving away from the source and building 'a little cabin in the woods' (Kunstler, 1996, p. 44). The problem with such a mentality is that it is fundamentally antisocial. Nevertheless, we continue to believe that as long as we create spatial barriers between land uses and classes of people, we can ignore the fundamental causes of our economic, social, and environmental problems. It is not surprising that places which grew out of this ideology failed to become the tight-knit communities they replaced.

Not only was suburbanization an attempt to physically recreate rural and small-town life, but it was also an attempt to preserve a set of values associated with an earlier time. The problem, as Warren (1978) argues, was that these older values had already changed. This final aspect of the Great Change affected patterns of community interaction in many ways. In his discussion of changing values, Warren (1978) lists several value changes that collectively represented a turn to greater rationalization: (i) gradual acceptance of governmental activity as a positive value in an increasing number of fields; (ii) gradual change from a moral to a causal interpretation of human behaviour; (iii) change in community approach to social problems from that of moral reform to that of planning; and (iv) a change of emphasis from work and production to enjoyment and consumption. Whether these value changes were a cause or consequence of the other aspects of the Great Change is not clear. What is clear, however, is that without this ideational component, none of the other changes could have taken root.

Great Change or Gradual Transformation?

Since it first appeared in 1963, Warren's Great Change thesis has been subjected to much criticism. The most important critique centres around his failure to examine community change in the proper historical context. Because he never provided a specific time line or description of community life before the Great Change, his theory is arguably rooted more firmly in assertion than empirical evidence. Take, for instance, the argument that as communities have become increasingly linked with larger systems beyond their borders, the locus of decision making has shifted to places outside the immediate locality. While this is an accurate description of life in most, if not all, contemporary communities, it is important to ask when the shift occurred. The historical evidence suggests that extensive vertical linkages began to be an important part of community life in the early 18th century:

> American regions, as soon as transport permitted, developed goods for export to markets thousands of miles away. The early town centers were located not only so as to serve the agricultural stratum but so as to serve the export of the region's staples.
>
> (North, 1955, p. 253)

In fact, many early towns were planned by urban entrepreneurs for the purpose of processing and transporting coal, minerals, forestry products, agricultural commodities and the products of other extractive industries. Control over these industries and the economic life of the local population was often located hundreds or thousands of miles away. Joseph Medill, a delegate to the Illinois Constitutional Convention of 1870, described this relationship well in his condemnation of Chicago's grain elevators and their owners:

> The fifty million bushels of grain that pass into and out of the city of Chicago per annum, are controlled by absolutely a few warehouse men and the officers of railways. They form the grand ring, that wrings the sweat and blood out of the producers of Illinois. There is no provision in the fundamental law standing between unrestricted avarice of monopoly and the common rights of the people; but the great, laborious, patient ox, the farmer, is bitten and bled, harassed and tortured, by these rapacious, blood sucking insects.
> (Medill, 1870, quoted in Cronon, 1991, p. 140)

Similar statements could have been made about the timber barons, railroads or coal companies. In short, the notion that there was once some golden era of community autonomy squares better with the narrative of rugged individualism than it does with available historical evidence.

A similar criticism can be made about the argument that local forms of solidarity and identity were undermined by the advances in transportation and communication that accompanied and facilitated vertical integration. While there is no doubt that cars, planes, telephones, fax machines and computers have liberated relationships from the bonds of locality and made possible spatially dispersed primary ties, it does not necessarily follow that local ties have withered. It is one thing to have increased access to the outside world; it is quite another to argue that the mere fact of this access has reduced the level and quality of interaction at the local level.

In one of the few studies to address this issue explicitly, Wellman (1979) investigated the strength of primary ties among the residents of the East York section of Toronto. Consistent with the notion that access to transportation and new communication technologies led to a decreasing reliance on local people and institutions, he found that 25% of East Yorkers' intimate ties were with people who lived outside the metropolitan area. On the other hand, approximately 25% of respondents' ties were to other East Yorkers, with an even higher percentage for ties to intimates who were not kin (Wellman, 1979, p. 1214).

In 1990, Wellman and Wortley restudied East York and found once again that reports of community demise were premature. Although the residents they interviewed did not describe their social networks in terms of the densely knit, *gemeinschaft*-like relationships that are often associated with our more nostalgic conceptions of community, neither did they describe their lives in terms of the alienation and isolation that the Great Change would suggest. In fact, most respondents reported that they relied heavily on local networks of family, friends, neighbours and co-workers for social support for emotional problems, homemaking chores and domestic crises (Wellman and Wortley, 1990, p. 584).

The Interactional Approach to Community

What do these critiques mean for the Great Change thesis and, in turn, for a definition of community? There is no doubt that communities have changed, but, as Kenneth Wilkinson (1991) pointed out, the conception of community that was supposedly lost during the Great Change has not existed for centuries anyway:

> What the sociology of community expresses mainly is a conception of community relevant to the Middle Ages and a lament that community thus conceived is being destroyed by long-distance communications, multi-site organizations, rationality of culture, and other modern trends. What is needed is a conception of community that recognizes its complexity.
> (Wilkinson, 1991, p. 7)

Warren, in the third edition of the *The Community in America* (1978, p. 417), admitted that it had become somewhat 'misleading to treat communities as concrete collectivities' and more theoretically and practically useful to conceptualize them in dynamic terms 'as a field of interaction'.

The interactional perspective, as this approach has come to be known, acknowledges the massive changes that have affected community life over the past century and poses the following question: Is there a key feature of local life that persists despite the transformations that have made previous conceptions of the community obsolete? The answer to this question is an unequivocal 'yes'. Even if the local community is not the integrated, holistic unit it might once have been, it is still the primary setting and mechanism for contact between the individual and society. Society is an abstraction that can never be experienced directly. The local community, by contrast, represents a tangible – albeit partial – manifestation of the larger social order (Konig, 1968). It is at this mesostructural level that most people meet their daily needs, and it is at least partially through the interactions which occur there that people develop a social definition of the self and beliefs about the way the larger society operates. Moreover, people who share a common territory inevitably interact with one another regardless of the extent to which they also participate in extra-local structures. From this perspective, then, social interaction is the essential element of community:

> Social interaction delineates a territory as the community locale; it provides the associations that comprise the local society; it gives direction to processes of collective action; and it is the source of community identity.
> (Wilkinson, 1991, p. 13)

Interaction also affects social behaviour in important ways. We behave and act purposively in response to the concept we have of our connections with others. This is a point Tonnies (1957) made long ago in his distinction between the natural will and the rational will. The natural will is impulsive; it is non-deliberative and non-calculative. In contrast, the rational will is deliberative; means and ends are considered and recognition is given to the necessity of suppressing impulses in order to attain goals. This distinction, perhaps because it is associated with the transition from a folk to a modern society, is often interpreted as characterizing two very different ways of being in relation to other people. A careful reading of Tonnies suggests that this is not the case. In all relationships the type of will varies – at times it is natural while at others it is quite rational. *Gemeinschaft* refers to those associations in which the natural will predominates. *Gessellschaft* refers to those characterized primarily by the rational will. However, *gemeinschaft* and *gessellschaft* are not polar opposites; elements of both are present in all relationships.

Herman Schmalenbach (1961) argued that the study of community required a third category of relationships (beyond *gemeinschaft* and *gessellschaft*) because of the use and misuse of Tonnies's concepts, especially *gemeinschaft*. In Schmalenbach's scheme, community is a natural state of being. The formative conditions of this state are not consciously experienced. Instead, they are experienced in an unreflective manner; the mere fact of human existence leads people into multiple and natural relationships with others. Taken as a whole, these relationships make up a common life. Community is experienced as pre-existing and is self-evident in people's behaviour. 'Recognition of community can arouse feeling, but community itself simply refers to the fact that one naturally is connected to other people' (Wilkinson, 1991, p. 16).

When community, as a state of being in relation to others, is consciously recognized and responded to emotionally, and that emotion is shared in social interaction, a new state emerges. This state is different from community, *gemeinschaft* or *gessellschaft*. Schmalenbach calls it *bund*, a term that can be loosely translated as communion. Communion is a conscious recognition of the natural bonds that exist between people; it is a celebration of community. Communion differs from *gessellschaft* because it lacks the crucial aspects of deliberation and calculation. It also differs from *gemeinschaft* because it is rooted in emotion; and, although communion depends on community, the opposite relation does not hold. For Schmalenbach, community exists whether it is emotionally celebrated or not.

Wilkinson (1991, pp. 16–17) draws on Schmalenbach's conception of community to indicate that it is a natural and ubiquitous phenomenon:

> It is natural because people, by the nature of being human, engage in social relationships with others on a continuing basis and they derive their social being and identities from social interaction. Community, likewise, is ubiquitous by virtue of the fact that all people engage in it almost all of the time, whether or not they recognize that fact. From the natural flow of interaction processes, community emerges, and the fact of its existence, whether or not celebrated in communion, affects the social processes that follow its emergence. Community, therefore, is a natural disposition among people who interact with one another on matters that comprise a common life.

People who share a common territory also inevitably interact with one another on place-specific matters. Interaction among community residents contains a special form of *gemeinschaft* that is characterized by a locality orientation. This locality orientation helps to create a generalized bond 'that cuts across and links special interest activities within the local territory' (Wilkinson, 1991, p. 37). In other words, locality-oriented interaction acts as a structure-building force that gives form to local social life. According to Wilkinson (1991, p. 39):

> Each actor has a real interest in the *local* aspects of local social life. This interest, which local residents have in common whether or not they experience it consciously, is pursued in social interaction and thus is shared.

Form is also given to local life as people organize themselves to accomplish tasks and pursue interests. From the interactional perspective, this organization is viewed in dynamic, processual terms. Rather than describing organized groupings as systems or subsystems, the interactional perspective describes them as unbounded fields of interaction. The community, in turn, is composed of several distinct social fields through which actors pursue or express particular interests. For instance, in most communities there is a social field composed of people and organizations whose main interest is economic development. Other fields focus on social services, health care, charity and so forth. Some of these fields are obviously more oriented toward local affairs than others. Those that are locality-oriented provide the raw material necessary to link separate fields into a holistic unit.

The mechanism that actually connects these multiple and different spheres of action is the community field. While the community field is similar in many respects to the other social fields in the community, it differs in one important respect. Rather than pursuing specific interests, as each social field does, the community field pursues the general community interest (Wilkinson, 1991, p. 90).

> The community field cuts across organized groups and across other interaction fields in a local population. It abstracts and combines the locality-relevant aspects of the specialized interest fields, and integrates the other fields into a generalized whole. It does this by creating and maintaining linkages among fields that otherwise are directed toward more limited interests. As this community field arises out of the various special interest fields in a locality, it in turn influences those special interest fields and asserts the community interest in the various spheres of local social activity.
>
> (Wilkinson, 1991, p. 36)

The actions that occur in the community field coordinate the more narrowly focused actions that occur in other social fields, and in the process bind them into a larger whole – albeit an unbounded, dynamic, emergent whole. The complex of interactions and actions that comprise community in turn contribute 'directly and positively to the social well-being of local residents' (Wilkinson, 1991, p. 67).

Conclusion

When it comes to understanding the relationship between social change and community, it is helpful to return to the idea that community is rooted in locality-relevant social interaction. To say that community is a natural and ubiquitous phenomenon rooted in locality-oriented interaction does not mean that social interaction must be based primarily on positive sentiment. Such a line of reasoning would be hopelessly unrealistic and would so narrow the range of interactional possibilities that the concept of community would become theoretically and practically meaningless. Clearly, a less restrictive approach is needed, one that recognizes and allows for a wide range of interaction. People interact with one another in all sorts of ways. What is important for the emergence of community is the interaction:

> Community simply depends on people interacting with one another ... Even as they are engaged in the most calculating of exchanges ... they engage simultaneously in *gemeinschaft*. Moreover, community entails squabbles and fights as well as cooperation and affectionate touches.
>
> (Wilkinson, 1991, p. 17)

In short, community depends on interaction – and when interaction is suppressed, community is limited (Luloff and Swanson, 1995; Bridger and Luloff, 1999, 2001). Thus, analyses of the relationship between community and social change must focus on the ways in which internal and external forces affect patterns of interaction.

Further, the level and extent of interaction among people who share a common territory can vary widely over time. This means that the extent of communityness among a local population can also vary substantially over time. An interactional approach to community and social change suggests that we focus on factors that suppress or facilitate the emergence of community. Understanding how stratification, racism, changes in the built environment, population growth and decline, and myriad other forces affect the extent and quality of interaction can provide us with important clues about what makes for healthy communities.

Notes

[1]Representative examples of this tradition include Hillery (1968), Parsons (1951), Loomis and Beegle (1957a), Loomis (1960), Sanders (1958), Bates and Bacon (1972), and Warren (1978). Human ecology, the major theoretical alternative to systems theory during this time period, also emphasized the patterned structure of relationships. In this sense, it is quite similar to systems theory. Compare, for instance, Amos Hawley's (1950) ecological definition of the community with Roland Warren's (1978) definition of the community as a social system. Hawley (1950, p. 180) defines the community as 'the structure of relationships through

which a localized population provides its daily requirements'. According to Warren (1978, p. 138), the community can be defined as 'that combination of social units and systems that perform the major social functions having locality relevance'. As these definitions illustrate, both the ecological approach and the systems approach highlight structure, order and organization in community life (Luloff, 1990).

Sublette, Kansas: Persistence and Change in Haskell County

Leonard Bloomquist, Duane Williams and Jeffrey C. Bridger

Introduction

Sublette, Kansas is located in rural Haskell County, approximately 200 miles west of Wichita in the south-western corner of the state. In 1940, Sublette was a community still reeling from the Great Depression and the drought and dust storms that ravaged the Plains throughout the 1930s. Like many communities in this part of the country, Sublette lost much of its population in the wake of repeated crop failures and the consequent economic decline. For these reasons, Sublette and its surrounding hinterland was seen as the most unstable community in the Rural Life Study series.

According to Earl Bell's original analysis, four key factors shaped community life in Haskell County: (i) the environment; (ii) federal government programmes; (iii) mechanization, commercialization and monoculturalization of agriculture; and (iv) human psychology. In June and July of 1965, Mays (1968) visited Haskell County to study the changes that had occurred over the 25 years since Bell's 1940 analysis. His examination found that the same forces remained important. Our analysis, based on a series of semi-structured key informant interviews as well as a 1993 representative sample survey of community residents, suggests that these factors continue to play an important role in economic and social affairs.

Environment: the Story of Irrigation Continues

In 1942, the factor most responsible for Haskell County's history of instability was what Bell called the capricious forces of nature, especially the unpredictable rainfall that could drench one field and completely miss another less than 5 miles away (Bell, 1942, p. 15). Moreover, in some years it rained as much as 20 inches, while in others the county recorded less than 5 inches. Not surprisingly, wheat harvests fluctuated wildly. In 1925, farmers averaged 20 bushels per acre; in 1938 they harvested just six (Bell, 1942, p. 23). This situation made for an extremely unstable economic base.

Although the weather patterns of Haskell County have remained capricious, cost-effective groundwater irrigation has made agriculture a more predictable undertaking. The development of pump engineering, availability of ground water from the Ogallala Aquifer, abundant reserves of natural gas and flat topology of the area (which permitted gravity flow application of the water) combined to facilitate the rapid spread of irrigated agricultural production after 1940. Indeed, as Mays (1968) indicated, 'By 1965, there were 263 wells irrigating approximately 204,000 acres' (p. 18). By the early 1970s, Haskell County was the most intensively irrigated county in Kansas.

In addition to increasing productivity and stabilizing the agricultural sector, irrigation also generated significant structural changes in the farm sector. Compared to both the US and Kansas, Haskell County has retained a higher proportion of its farms and experienced less growth in farm size over the past 50 years, although its farm production has diversified and become much more intensive. Energy and hired labour costs have increased significantly, and are considerably above national and state averages. Off-farm employment is much less prevalent in Haskell County than in either the US or other areas of Kansas.

Fig. 3.1. Sublette, Kansas, 1942. Grain was often stored in small on-farm storage bins in the 1940s.

Much of the economic development associated with irrigation is reflective of the large increases in feed grain production, which in turn attracted cattle feedlots to the area (Williams, 1995). The concentration of feedlots contributed to the opening of several large meat-packing plants in the region. In the early 1980s, Iowa Beef Packers (IBP) opened the world's largest beef-packing plant in Garden City. Packing plants in Finney, Ford and Seward counties slaughter approximately 20,000 head of cattle each day. Haskell County is centrally located in the midst of the three communities where these facilities have developed (see Fig. 3.2).

The development of feedlots and packing plants in the region has substantially altered the region's socio-economic characteristics. In 1990, agriculture accounted for one-third of the employment in the county – a level ten times higher than for the USA as a whole. Between 1970 and 1990, Haskell County's agricultural employment increased substantially, and ran counter to the decline experienced overall in Kansas. While grain production was once the primary form of agricultural production, most farm income now comes from the sale of cattle. In 1991, for instance, over 83% of Haskell County's gross farm sales came from the sale of livestock.

As agriculture became a more mature and stable industry, the dramatic fluctuations in migration patterns that characterized the county in the early 1900s ceased. In contrast to the population losses that characterized most agriculturally dependent counties in rural areas of Kansas and the Great Plains, Haskell County has grown consistently since 1940 from slightly over 2000 residents at that time to 3886 in 1990. The average length of residency in Haskell County now exceeds both national and state averages.

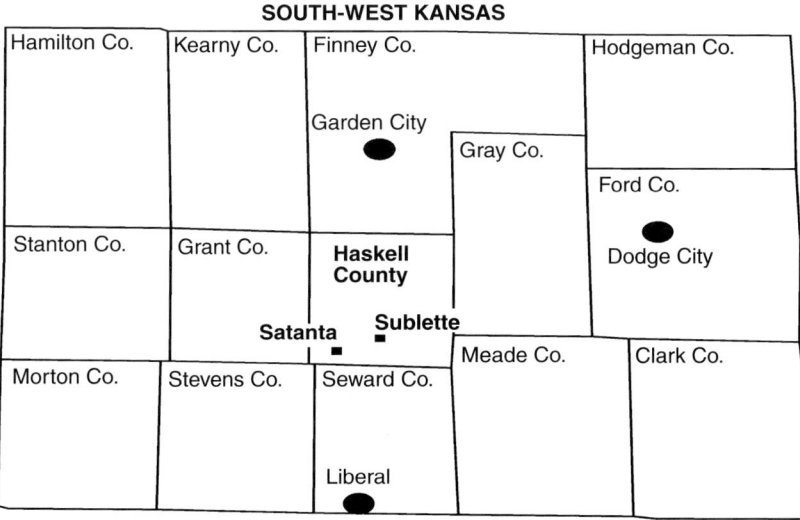

Fig. 3.2. Haskell County and surrounding counties.

With the growth of the meat-packing industry, ethnic diversity has also increased. Latinos, especially, have migrated to the area over the past 20 years to work in the packing plants, and are now an important cultural presence. Census figures from 1990 revealed that approximately 13% of permanent residents in Haskell County were of Hispanic origin.

Whether the population growth that was made possible by irrigation will continue is open to question. Since the mid-1970s, the future of irrigation has become problematic. The Ogallala is the fastest declining aquifer in the world, with the rate of withdrawal over natural replenishment almost equal to the flow of the Colorado River (Reisner, 1993, p. 10). In 1991 and 1992, the decline of Haskell County's water table was the most severe in the south-west region of Kansas. Many of the older wells are reaching the limit of their expected useful life. Moreover, the cost of the natural gas that powers the wells has fluctuated widely with deregulation. Together, these factors are beginning to threaten the long-term viability of irrigation-extensive agriculture. Although new irrigation technologies may help to alleviate the growing water shortage, it is too early to tell whether they will provide a long-term solution that is both economically and environmentally sustainable.

The increased role of the state in water issues will also affect the future of irrigation. The local Groundwater Management District, a quasi-public institution made up of water users, of which Haskell County is a part, is implementing a system to meter the use of water. While most farm producers understand the fragile nature of the aquifer's water resource, and most support conservation and the wise use of water, many question the effectiveness of the metering system and resent the increased presence of the state. As one farmer explained, 'The increased role of government in the conservation of water is not the answer – metering, most farmers are against it.'

Should irrigated agriculture not be viable, Haskell County's natural gas resources might help to buffer some of the resulting economic impact. The county sits atop the Hugoton gas field, one of the largest known natural gas reserves in the world. However, like most energy reserves, the field is controlled by large corporations and is affected by an array of external forces. In the 1990s, rapidly changing global markets, federal deregulation, technological advances, and other factors negatively affected the local energy sector.

Government: a Change in Local Attitudes Toward Outside Governmental Presence

In 1940, the federal government developed an array of programmes designed to support agriculture and to help communities cope with hardships brought on by the Great Depression (Bell, 1942). By the time Mays

Fig. 3.3. Sublette, Kansas, 1995. Most grain was stored in Sublette's grain elevator until ready for shipping.

visited the county in 1965, such programmes had declined in number and were primarily limited to agriculture and public schools. However, during interviews conducted in 1993, several residents indicated that state and federal government programmes had re-emerged and were once again playing significant roles in their community.

Since then, people's attitudes toward outside governmental actions and involvement appear to have changed dramatically. Many people in Haskell County, particularly those who lived through the 'dust bowl' and Great Depression era, recalled the important role that federal government programmes played in the history of their community. They acknowledged that the farm programmes and other 'New Deal' programmes of the 1930s brought stability to the community. At the same time, dissatisfaction with government involvement in local affairs was widely expressed. Some people noted that this reflects changing attitudes in society in general and a growing tension between higher expectations about government support and the desire to preserve individual freedom. Others discussed the loss of rural political influence and an increasingly urban bias in public policy. One resident summarized these views when he commented that:

> ... declining political influence is frustrating farmers. Some older farmers dislike the ASCS–SCS [Agricultural Stabilization and Conservation Service, Soil Conservation Service]. Younger farmers are becoming dependent, but often retain or reflect the older farmers' attitudes.

Another said: 'Schools have seen increasing federal and state involvement, [we could be moving toward] a state–national curriculum.' Others felt that local autonomy was being eroded as the federal government has grown in size and power: 'With government programs, the structure of local input continues to exist, but little areas are left for real local decision making.'

Results from our 1993 community survey corroborated the concerns expressed in key informant interviews. When asked about problems facing their community, the two most frequently mentioned issues were increasing taxation and the loss of local control to the state and federal governments. The sense of declining local control was most acutely felt in the educational arena.

For the south-west region of Kansas, which includes Haskell County, the combination of complex social issues, court mandates, increasing regulations that carry no fiscal support, the relative decline of rural representation, the region's abundant natural resources and a relatively high degree of wealth relative to other rural areas has generated an atmosphere of suspicion toward the role or presence of federal and state governments. The need for, or desirability of, many regulations is not the issue. Like people everywhere, Haskell County residents want safe water, a clean environment, sound financial institutions and fair labour practices. What they resent is the rigid application of regulations they view as developed to meet the desires and problems of urban areas. They are also aware of diminished state and federal assistance to help local areas implement many mandates.

On the basis of both the key informant interviews and the community survey, it was clear that Haskell County residents felt that the government's presence was increasing, and they sensed that more local resources were being drawn away from the community. Many perceived a division

Fig. 3.4. Sublette, Kansas, 1942. A typical farmhouse in 1940.

between eastern and western Kansas, one based on a rural/urban distinction. Recent actions by some in the south-west region of Kansas to secede from the state provided a clear indication of the seriousness of these concerns.

Not all state and federal programmes were viewed negatively in Haskell County, however. Programmes that provided support to senior citizens and the poor, and local economic development efforts were generally viewed positively. Not surprisingly, each of these programmes reflected a transfer of resources into the local community.

A content analysis of the local newspaper, while supporting the perception of a relative decline in the presence of outside governmental units between 1940 and 1965, did not confirm the re-emergence of these activities in 1993. In 1940, as Bell observed, federal programmes were pervasive. Nearly half of the relevant articles (within the content analysis coding system) were about extra-local agencies and activities. In 1965, the largest share of these articles was about local community activity. Just as Mays (1968) had reported, economic stability had allowed the features of a more stable and complete community to emerge in Haskell County. The dominance of local community activity continued in 1993. Beyond the fact that these articles still represented the majority view, a clear pattern of increased community development activity was visible.

There are several possible explanations for the discrepancy between the results of the semi-structured interviews, the survey of Haskell County residents and the results of the content analysis of the local newspaper. One is that federal and state government activities have become such an integral feature of daily life that these actions, while having a significant impact upon the local community, are no longer considered 'news'. Another possible explanation is that the conflict between the local community and extragovernmental agencies is perceived by residents to be more intrusive than in fact it is. Still another possible factor is that the growth of regional newspapers has changed the scope of news coverage in the local newspaper since 1940. Many rural weekly papers focus almost exclusively on local events, leaving most of the state, national and international news coverage to the region's daily newspaper. In reality, the situation in Haskell County in 1993 is probably a combination of all of these factors.

Agriculture: the Evolution into Agri-business Continues

In 1940, Bell (1942) found that Haskell County contained a commercialized agricultural monoculture focused upon wheat production. Farming was highly mechanized and dependent upon the 'cash' economy. The frailty of this economy became painfully clear after the price of wheat fell in 1931 and the rains stopped in 1932 (Edwards, 1939; Bell, 1942). In 1965, Mays (1968, pp. 35–36) noted that agriculture in the area was becoming 'agri-business'. The rising capital cost and year-round production activity of irrigation were key aspects of this change.

These trends have continued since 1965. Massive capital investments, increased mechanization, a greater use of hired labour, changing technology (i.e. new crop hybrids and the increased use of chemicals and fertilizer), deregulation of energy inputs, the rising importance of successful marketing strategies and the increasing scope of government regulations all have changed dramatically the business of farming. Today, management skills have replaced production capabilities as the defining characteristic of successful farming. Successful managers are those who are able to deal with the rising complexity of farming. According to several residents, the ability to manage the increasing complexity in farming is regarded as 'progressiveness'.

While irrigated production has increased the intensity and complexity of farming in Haskell County, it has also, until recently, allowed the county to buck the national and state trend of declining farm numbers and increasing farm size. Largely because the labour requirements of flood irrigation place a significant barrier to increasing economies of scale, Haskell County experienced growth in farm numbers and a decline in average farm size during the height of irrigation development. However, the emerging problems in irrigation (described above) in combination with the national farm crisis of the mid-1980s eliminated the county's ability to continue to buck the prevailing trends. Over the past 10–15 years, Haskell County experienced a drop in farm numbers and an increase in farm size. From 1987 to 1992, the rate of farm loss and the rate of increase in farm size in Haskell County exceeded both the national and Kansas averages.

Fig. 3.5. Sublette, Kansas, 1995. A typical farmhouse today.

Federal farm programmes remain an important aspect of farming in Haskell County. In his original study, Bell reported extensively on the role and importance of the newly developed farm programmes, and the attitudes of the community's farmers regarding them. In 1940, farmers expressed a reluctant acceptance of federal farm programmes. They recognized their dependence upon the programmes for survival, but regretted programme requirements, which they felt diminished their independence. These same sentiments existed in 1993. While local farm operators felt that 'participation was a must', they also believed that 'the programmes had become a bureaucratic nightmare', as one respondent explained. Government regulations had escalated, and each regulation had added to the amount and complexity of paperwork faced by farm operators. There was a sense that many of these regulations were generated by non-farming, special-interest groups. Such regulations appear to farmers to be formulated in a vacuum, ignoring regional differences. While frustrated by increasing regulations, some farm producers also recognized that they needed assistance in managing the increasing sophistication of new technology. One farmer suggested that 'regulations may be the best means to ensure safety and compliance'.

The increasing complexity of farming, and the concomitant emphasis upon management, represents a dramatic change in this occupation. While the physical stress associated with farming has declined, mental stress has increased. Air-conditioned tractor cabs and hydraulic-powered equipment have reduced the physical stamina that farming once required. The 'sweat of the brow' is now more likely to occur as the farm operator slides behind his desk to tackle increasingly complex paperwork. Farm programme records, financial statements, marketing transactions, documentation of compliance with government regulations, and income tax documentation and preparation all make farming an increasingly stressful occupation. Given the amount of capital required in today's farm operations and the recent changes in lending policies, financial management is both one of the most critical, and most stressful, aspects of farming.

Human Psychology and Social Ties: Contradictory New Features Emerge

In 1940, Bell reported that the wide variance between the years of bounty and the years of absolute crop failure had a strong psychological effect upon those living in Haskell County: 'It has helped to develop the gambler's psychology, so noticeable to outsiders' (1942, p. 42). To many of the farmers he interviewed, a successful crop was simply a matter of luck. For instance, when asked to explain why he was able to raise a feed crop on land that wasn't as productive as his neighbours', one young farmer replied: 'I just hit it lucky' (Bell, 1942, p. 42). By 1965, Mays (1968) reported that business management skills had replaced luck as the key to success in the psychology of the community.

This was also true in 1993. Progressiveness, the ability to manage successfully the increasing complexity of farming, was now viewed as the most important ingredient for economic success. However, people's attitudes also reflected other changes. Increased length of residence and the accumulation of wealth resulted in greater class stratification in Haskell County. The top social class consisted of wealthy farm families who could trace their roots back to the pioneers who settled the region. As might be expected, older individuals make up the majority of this group. Their primary concern is stability, and they act to maintain a vision of the community rooted in the past. Because of their wealth and power, these individuals have been very effective in promoting this perspective as the dominant ideology in the community. However, many county residents hold somewhat contradictory attitudes. They espouse progressiveness in business affairs but hold to a nostalgic view of community social life.

With the decline in the energy sector in the early 1980s and the agricultural crisis of the mid-1980s, the no-change ideology of the older and established leadership began to be questioned. A new group of leaders, largely composed of professionals and managers who were not life-long members of the community, emerged. By the 1990s, these leaders were actively confronting the established leadership as they attempted to promote and legitimize a pro-growth ideology. The struggle has created tensions between those who want to maintain the status quo and those who would like to see the progressiveness associated with successful farming become an important force in other areas of community life.

Despite the changes that have occurred since 1940, Sublette and Haskell County have retained many of the traditional community characteristics evident in Bell's initial study. Survey results indicate that community attachment is quite high across this county. Family, friendship and economic ties contribute to this strong level of attachment. Most Haskell County residents reported that they were very interested in local community happenings and that they would be sorry if they had to leave the area. Familiarity and kinship networks were quite extensive. Community and organizational participation was high; most residents were members of at least one community organization.

In 1965 Mays (1968, p. 69) found that the family was the primary informal social unit and the school was the dominant social institution in Haskell County. Although rising individualism and changing attitudes about the care of the elderly have reshaped the family unit, it continues to be the primary informal social unit in Haskell County. The school also continues to be one of the most important social institutions. As was the case in 1940 when Bell visited the county, the school continues to fulfil multiple functions for the community. In addition to its obvious role in educating and socializing children, the school districts are also the largest employers in the county. In addition, extra-curricular school activities continue to provide the largest proportion of local recreation/entertainment. School activities remain the most important locus of local social interaction in Haskell County.

From 1940 to 1965 the churches played a very traditional role in the community. By the 1990s, the churches were identified as the third most important element of communal relations (following family and schools). However, in contrast with findings from the 1940 and 1965 studies, the churches had become much more visible and active in community affairs in the 1990s.

Another important aspect of social institutions and social interaction was the changing nature of community leadership. In contrast to 1940, when men avoided formal leadership positions and women actively sought such opportunities, in 1965, men were more active than women in formal leadership roles (Mays, 1968). Leadership through public office had replaced the informal leadership by men in 1940. In 1993, the role of women in Haskell County was closer to Bell's (1942) description, with a number of women filling formal leadership positions.

A Theoretical Evaluation of Change in Haskell County

In this section, we integrate the information gained from the three community case studies of Haskell County with sociological theories of community organization and change. We begin with an interpretation of the findings utilizing Roland Warren's (1978) functional approach. Information from the three community case studies of Haskell County is evaluated for evidence of Warren's (1978) 'Great Change' processes. This evidence is then used to place the Haskell County community within Warren's multidimensional framework for each of the three points in time. A comparison of the community's location within this framework across these time periods provides the means for assessing Warren's 'Great Change' paradigm. Anomalies between observed patterns of change and the elements of Warren's theory are then examined utilizing the interactional perspective elaborated by Kenneth Wilkinson (1991).

The Community's Location Within Warren's Multidimensional Framework

Haskell County in 1940

While some evidence of the traditional community was found in 1940, overall, Bell's study suggested that Haskell County's communities contained few traditional characteristics. They had little *local autonomy*. Their *production–distribution–consumption* functions had an outside focus. Agriculture dominated Haskell County's economy. Its farmers were dependent upon the cash market. They sold everything they produced and bought everything they consumed. Much of what they consumed, both agricultural inputs and family needs, was produced outside the local community. The capricious forces of nature provided a very unstable economic environment. Economic instability, in turn, impeded commercial and social development. The availability of automobile transportation greatly expanded travel to obtain goods and services.

Socialization was the one facet of the locality relevant functions that was locally based. While the community's history of significant population migration had limited the formation of shared values and practices, the homogeneous composition of the county's population did provide a common set of beliefs and norms. The family was the primary informal social unit and the schools were the dominant formal social institution. Some of the socialization functions of primary groups, i.e. the family and neighbourhood, had been transferred to the school. Social interaction was sexually segregated. Men interacted informally, focusing upon practical economic functions. Women interacted both informally and formally. Most of the socialization that occurred outside the home and school was through the social interaction of women.

Local *social control* was more limited. The role of local government in setting and enforcing community standards fell beneath the weight of state and federal relief programmes designed to aid farmers through the dust-bowl-era drought and assist the overall community through the country's economic depression. However, the community's family focus provided strong support for maintaining norms and values. Churches, although lacking strong participation and financial support, continued to play a significant role in enforcing individual conformity.

These conditions also affected *social participation*. The spirit of neighbourhood cooperation and interaction, prevalent during pioneer settlement, was replaced with a focus within the extended family. Family and school activities were the two primary forms of social participation. School activities dominated the social calendar. Social participation through church activities was relatively limited. Social interaction by men was predominately unorganized. Although women had a well-established array of clubs and organizations, men typically avoided formal public leadership roles.

The combination of the drought and economic depression exhausted local resources for *mutual support*. Haskell County, like many other American communities at that time, was dependent upon state and federal assistance.

There was some *variance in the community's service areas*. The service area of local schools and churches generally coincided with the geographic location of the community's residents. However, with the increased availability of automobile transportation, local residents frequently shopped for commercial goods and services outside the county. The commercial service area had grown to approximately 100 miles in diameter. Moreover, local governmental services were primarily an outreach of state and federal programmes.

Residents' *psychological identification with their locality* was rather limited, as Bell (1942) found only one example of a socially cohesive community in the county. Historically, the county had a highly transitory population. The limited desire for land ownership among farm operators was an indication of the weak identification with the locality. Community development had been stunted by instability. The dominant role played by state and federal programmes had a negative impact upon residents' identification with the local community.

The community's vertical pattern was stronger than its *horizontal pattern*. Instability limited social and community development. The strongest institutional interaction of local governmental units was with state and federal programmes.

Haskell County in 1965

By 1965, traditional community characteristics in Haskell County had become noticeably stronger than in 1940. The community's *local autonomy* was strengthened. Some of the community's *production–distribution–consumption* functions had become more inwardly focused. Irrigation had provided agricultural stability. This agricultural stability in turn fostered economic and social stability. Local commercial activities expanded. Trade patterns shifted towards a preference for local consumption. Industrial development helped to diversify the local economy. Nevertheless, linkages to extracommunity systems remained strong. Farming had become more commercialized and farmers were still dependent upon the cash market. Federal farm programmes were still very important. Although diversification had shattered the wheat monoculture that had dominated agriculture for years, Haskell County farms and farm families continued to purchase most of their products and services.

Socialization retained, and perhaps strengthened, its local focus in the 25 years following 1940. Economic and social stability, as well as the continued homogeneous composition of the population, strengthened local norms and values. Rising individualism had not shaken the family from its position as the primary informal social unit. Socially, the role of the school had expanded, and its stature as the dominant formal social institution had increased. The social interaction of men in formal activities increased as well.

Local *social control* was markedly improved in 1965. The role of local government in helping to set and enforce community standards was

Fig. 3.6. Sublette, Kansas, 1942. The few farms with irrigation in 1940 all used flood irrigation.

enhanced by economic and social stability. Despite some changes in family values (i.e. rising individualism and changing attitudes about mutual family support), the family continued to be a strong force for maintaining social norms. Participation in and financial support of local churches improved; churches continued to help set and enforce community norms.

Social participation had also increased in 1965. Generally, informal visiting remained limited to family members. The school continued to be the primary vehicle for community interaction. Church activities expanded, but they remained largely outside of community service endeavours. Although men still did not actively participate in most social organization, their involvement in formal public leadership roles was dramatically greater than in 1940.

Overall, the local provision of *mutual support* had improved by 1965. Economic growth and relative stability had allowed the local government to re-establish its role in providing local support. Local administration of welfare programmes was cited for its forward-looking approach. Although the family was still an important source of mutual support among its members, rising individualism and a shift in attitude about its role in providing support had occurred. In particular, elder care was viewed by many as a public responsibility supported by taxes.

The *coincidence of community service areas* was also greater in 1965. Local school and church service areas continued to coincide with the location of residents. Increased commercial development helped to foster a greater propensity for shopping locally. The service area of local businesses differed less than it did in 1940. Local governmental services, which also coincided with other service areas, were more significant.

Residents' *psychological identification with their locality* had become much stronger in 1965 than in 1940. Economic and social stability had helped promote the development of community. A larger proportion of the population had lived in the community for a longer period of time. The commercial and governmental sectors had been strengthened. There was greater public involvement by men. Each of these factors strengthened psychological identification with locality. Another indication of a stronger attachment to the locality was that, in contrast to 1940, a majority of the farm operators preferred land ownership.

Finally, the *horizontal pattern* of the community was also stronger in 1965 than in 1940. The number and strength of local units had improved significantly with economic stability. However, agriculture was still strongly linked with federal programmes, and industrial development had been largely driven by outside interests. Otherwise, the linkage among the various local social institutions was stronger than their ties with extra-community institutions.

Haskell County in 1993

By 1993, the community's *local autonomy*, as measured by Warren's (1978) five locality relevant functions, was being pulled in different directions. The community's *production–distribution–consumption* functions were more outwardly focused than in 1965. The local economy had not diversified significantly beyond the agriculture and energy industries. However, the increased vertical linkages of agriculture and the emergence of a cattle complex, i.e. feedlots and packing plants, had increased the division of labour. Over time, the level of outside ownership within key economic sectors, such as the feedlots, packing plants, agricultural services, and the energy production and service sector, had grown. Regional, national and international control of these enterprises had clearly increased Haskell County's systemic relationships with the larger society. Further, commuting to reach employment outside the county increased.

With the depletion of the aquifer and the ageing of the irrigation infrastructure, irrigation and agriculture were less stable by 1993. The energy industry was in a down cycle. These conditions affected economic and social stability. The mix of local commercial activities decreased while the effects of mass merchandising increased. Trade patterns were shifting away from local consumption. On the other hand, local community economic development efforts directed at diversifying the local economy and improving commercial activity increased substantially.

Farming also became more commercialized and farmers were still producing for the cash market in 1993. Crop production became increasingly dependent on government programmes, and this limited the diversity of agricultural production. Haskell County farms and farm families continued to purchase most of their products and services.

Socialization retained, and perhaps strengthened, its local focus in 1993. The family maintained its position as the primary informal social unit. Familiarity and kinship networks were quite extensive in Haskell County. The role of the school in socialization continued to expand, and was still the dominant formal social institution in the community. Overall, local interest and support of the schools remained high. The churches became much more visible and active in community affairs. The role played by men in formal leadership positions continued to expand.

However, changes in socialization were evident in 1993. The decline in economic and social stability and the increased diversity of the population began to reshape local norms and values. The rising participation of women in the paid workforce, though less than in other places, modified family structure. Technological changes, particularly in communications and entertainment, were also influencing the patterns of social interaction.

In 1993, local *social control* had diminished, in perception if not in reality. The role of outside government had increased. However, the family continued to be a strong force for maintaining social norms. Participation in and financial support of local churches continued to be strong, and the churches were actively trying re-establish and enforce community norms.

Social participation continued to increase after 1965. In 1993, family interaction was still the most common aspect of informal visiting. Increasing length of residency fostered higher social and economic investments by most residents in the community. These investments, in turn, stimulated social participation. The school, as in 1965, continued to be the primary vehicle for community interaction. Church activities continued to expand, and the churches were more involved in community service endeavours. Participation by men in formal public leadership roles continued, though women were also actively involved in a number of public positions.

Some emerging characteristics of the community in 1993, however, could be indicative of negative trends with respect to local social participation. These included increased commuting for employment, growing participation of women in the paid workforce, and the increasing time required for irrigated agricultural production. In addition, technological changes, which foster contact with the broader society and support individualistic forms of recreation and entertainment, acted to reduce local social interaction and participation.

The local provision of *mutual support* lessened between 1965 and 1993. Long-term economic growth and general prosperity desensitized the community to needy local families. Within the community's culture – a culture based upon economic rationality – the prevailing attitude correlated hard work, material success and moral correctness. The provision of aid was viewed largely as the responsibility of the government.

The *coincidence of community service areas* also had lessened between 1965 and 1993. Local school and church service areas continued to coincide with the location of residents. However, changes in the mix of commercial activity and the rise of mass merchandising helped to foster a weaker propensity for shopping locally. The service area of local businesses differed more than it did in 1965. Out-commuting for employment also affected the location where people acquired goods/services. Finally, local governmental activities were increasingly tied to outside governmental actions.

Residents' *psychological identification with their locality* remained very strong in 1993. The economic and social stability that characterized most of the period between 1965 and 1993 promoted the development of community ties. The average length of residence in the local area increased during this period, and survey-based measures of community attachment and participation in community affairs were strong. Moreover, networks of family and friendship relationships were extensive. As a result, residents had significant social and economic ties to their community.

Interestingly, identification with the local community remained strong despite the presence of elements of the Great Change that, theoretically, should have lessened such feelings. These elements included the re-emergence of outside governmental forces, technological changes in the areas of communications and entertainment, out-commuting for employment and increased ethnic diversity and social stratification.

The *horizontal pattern* of the community had declined by 1993. Although the number and social interaction of local units was still significant, the emergence of mass society was discernable, particularly in regard to changes in communication technologies. Satellite dishes, cable television, videos and regional newspapers drew people's attention outward. Agriculture was even more strongly linked with federal programmes. Other federal initiatives were also influencing community activities. State government became more involved in the local schools. Recent economic development efforts were initiated as part of a state programme that focused on generating a more regional perspective.

An Evaluation of Warren's Great Change Paradigm

The community characteristics and trends outlined above were evaluated against Warren's (1978) discussion of the Great Change processes. Both supporting and conflicting evidence of these processes of change were discovered. This evidence was used to place the Haskell County community within Warren's multidimensional framework for each of the three points in time (see Fig. 3.7).

Based upon this comparison, two significant anomalies with respect to the Great Change paradigm were discovered. These anomalies were condi-

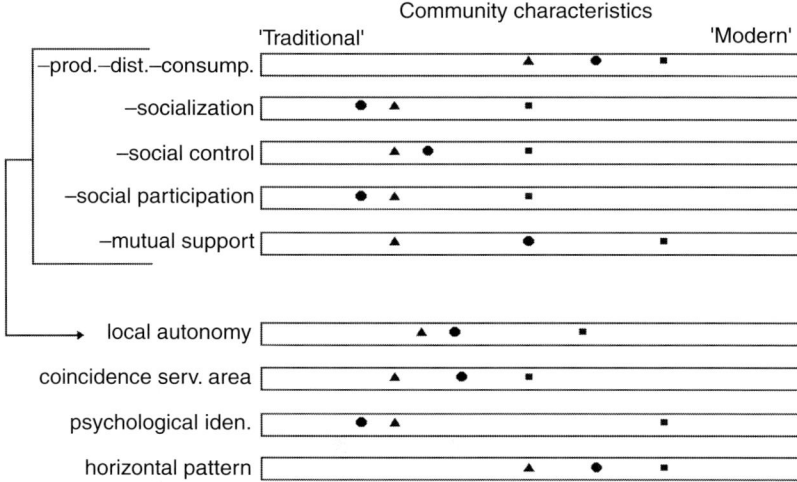

Fig. 3.7. Warren's (1978) multidimensional fields. ■, Haskell County in 1940; ▲, Haskell County in 1965; ●, Haskell County in 1993.

tions in conflict with Warren's position that each of the seven processes of the Great Change would move communities in the same direction on each of the four key dimensions of the community framework, i.e. away from the traditional conception of community toward a more modern conception.

The first anomaly was an increase in traditional community characteristics between 1940 and 1965. Haskell County became more of a traditional community on each of the four dimensions of Warren's (1978) multidimensional framework. These results ran directly counter to the changes proposed by Warren's model.

The second anomaly concerns the changes between 1965 and 1993. During this period, the overall changes in Haskell County's community characteristics were more consistent with Warren's (1978) Great Change paradigm, as decline occurred in traditional community characteristics in three of the four dimensions. However, Warren's proposition held that there would be a decline in all four dimensions. This did not occur, as residents' psychological identification with their community seems to have increased between 1965 and 1993. In addition, although the dimension evaluating local autonomy had experienced a decline, there were elements of the five locality-relevant functions that remained strong, or perhaps even increased (social participation increased and socialization was the same, or perhaps strengthened). In the next section, these anomalies will be examined utilizing the interactional conception of community (Wilkinson, 1991).

Fig. 3.8. Sublette, Kansas, 1995. Pivot irrigation systems were being installed in 1995.

The Interactional Perspective

Utilizing an interactional perspective, one could argue that the instability of Haskell County prior to the development of irrigation, especially the severity of conditions during the Great Depression and dust-bowl period, had drastically limited the extent to which the local territory could support a local society. During Bell's (1942) visit to Haskell County in 1940, the primary focus of people's time and energy was directed toward the struggle for survival. As economic conditions improved, and as irrigation provided stability in the context of the capricious forces of nature, the time and energy required to meet lower-order needs lessened. A lower burden meant that people could pursue higher-order needs and the community had the opportunity to emerge (see Wilkinson, 1991, p. 68). By 1965, Haskell County's local ecology was able to provide the forms of social organization required to meet daily needs. It contained a social organization that afforded a holistic interactional structure that provided for a more or less complete array of opportunities for association. Community action, representing the bond of local solidarity, was clearly evident.

In 1993, a re-emerging instability, an increased presence of mass society and the increased role of outside governments had a negative impact upon Haskell County's traditional community characteristics. According to the interactional perspective, such changes affect the emergence of community to the degree that they lessen locality-relevant social interaction.

This may help to explain why not all of the indicators represented by Warren's notion of the 'Great Change' were moving in the same direction. The continued strength of socialization and social participation as locality-relevant functions and the increase in residents' psychological identification with their locality indicated that local social interaction continued to support the natural emergence of community. Clearly, the level of economic instability in 1993 was a mere fraction of that present in 1940. While growth and prosperity in Haskell County may have faltered, most residents were not consumed by the struggle for survival, as was the case in 1940. Haskell County's local ecology continued to provide a social life that met a majority of its residents' daily needs. It still contained a social organization that afforded a holistic interactional structure of a more or less complete common life. Finally, the bond of local solidarity continued to be readily apparent through community action.

The interactional perspective provides insight into how changes in stability impact upon communal relations. Increased evidence of traditional community characteristics in Haskell County between 1940 and 1965 (within Warren's (1978) multidimensional framework) are consistent with concepts of the interactional perspective. The failure of processes representative of the Great Change to move all indicators in the same direction, towards a decline in traditional community characteristics, can also be explained from the interactional perspective. In short, even while some dimensions of the local social system appear to have eroded or become less relevant over time as extra-local linkages have expanded, other dimensions that reflect the interactional processes that lie at the heart of community have persisted.

The Future of Haskell County

Despite being characterized as extremely unstable at the time of the original Rural Studies series investigation in 1940, Sublette and Haskell County have exhibited patterns of change that, in many ways, run counter to the trends evident throughout much of rural America. Increased economic stabilization and the persistence, and even strengthening, of community social structures have occurred in the years since Bell's initial study.

There is substantial reason to anticipate that changes will continue to define this community setting in the coming years. This is due in part to the fact that water remains, in many ways, the defining resource of Haskell County. From the settlement of the county in the late 1880s through the time of Bell's visit in 1940, the capricious rainfall in the region had created a history of economic and social instability. By 1965, technological changes had unlocked the vast resources of the region – in particular the ground water in the Ogallala Aquifer. Irrigation became the foundation of economic and social stability. However, at the time of our study in 1993, the aquifer was in decline. This decline represents a fundamental challenge that will almost certainly shape the future of Sublette and Haskell County. The declining water table is already contributing to changes in agricultural practices. It also has caused some Haskell County residents to speculate that development of the community might have reached its peak. While some residents forecast the community's decline, others pointed to a need for economic development and diversification so that agricultural declines do not contribute to instability and deterioration.

When assessing the challenges that lie before them, many Haskell County residents recalled the accomplishments of past generations, and recounted tales of their community's rise from the desperate conditions of the 'dust bowl' and Great Depression. Drawing upon this heritage, many Haskell County residents have a guarded confidence that the community can, and will, meet the challenges it faces. While the lack of water may be re-emerging as a force to be reckoned with, the development of new technologies, such as centre pivot irrigation systems to conserve water may allow for continued prosperity. More importantly, the emergence of strong community ties and localized social actions in support of community goals and needs suggests that such challenges will be met by substantial local effort to maintain and sustain the community.

Irwin, Iowa: Persistence and Change in Shelby County

Eric O. Hoiberg

Introduction

Irwin is a small community in Shelby County, located in west-central Iowa. At the time of the original United States Department of Agriculture (USDA) Rural Life Studies, Irwin was characterized as occupying a midpoint on the instability–stability continuum – it was a community that had seen substantial change, but because of the dominant influence of agriculture, had remained relatively stable over time (Moe and Taylor, 1942).

Irwin was selected for the USDA study because it was believed to be a typical example of a Corn Belt community. Any number of other midwestern communities could have been chosen as representative of this community type, but Irwin was selected, at least in part, because one of the study's authors was a Shelby County native and thus generally familiar with the area's social and physical landscape. A more substantive reason for selecting Irwin was that its population in 1940 was relatively diverse. Unlike many of the other study communities, there was no dominant ethnic or religious group in the community.

The authors of the original study highlighted the levelling influence of a common ecological base and pointed to the fact that earlier distinctive social groups had been 'eliminated in the melting pot of similar tasks, neighbourhood cooperation in some of these tasks, common economic necessities and purposes' (Moe and Taylor, 1942, p. 90). Thus, Irwin was seen as a distinctively 'American' community, the cultural life of which was relatively settled and stable and was 'conditioned by, if not tuned to, the cycles and rhythms of the type of farming that prevails' (Moe and Taylor, 1942, p. 91).

Irwin is located in the north-east quadrant of Shelby County, situated in the second tier of counties east of the Missouri River and roughly halfway between the borders of Missouri to the south and Minnesota to the north. The sociological community is generally conceived as including four townships – Douglas, Polk, Greeley and Jefferson – with the village of Irwin physically located in the latter two administrative units.

The Study

This chapter centres on two images of Irwin, Iowa. The first is evoked by interviews with a number of older local residents who were in their prime in 1940 when a young rural sociologist named Ed Moe began his study of a typical midwestern agricultural community. The other image is drawn from Irwin's present and future, as seen through the eyes of other younger residents. Each of these group's visions transcend a particular historical period: the younger generation has a keen sense of the present and future while the older generation's views of contemporary Irwin are substantially influenced by an appreciation of the village's local culture and history (Fig. 4.1).

The observations reported in this chapter support a focused analysis that views Irwin as a community deeply impacted by dramatically changing agricultural and demographic patterns. It focuses on the adaptations made in the local institutional structure over the past 50 years and the impact of these changes on community attachment and local participation.

Fig. 4.1. Irwin, Iowa, in 1941. This image shows the Irwin business district in 1941.

Fieldwork for this study began in the fall of 1988 when the author was on sabbatical leave from the Department of Sociology at Iowa State University. Numerous trips were made to Irwin during an intensive 6-month period, followed by periodic visits over the next 8 years. In-depth interviews, attendance at community events and meetings, analysis of historical documents and artefacts, and casual observations contributed data to this study. Also, in the spring of 1993, a survey was sent to all Irwin households and to all open-country households in the four townships surrounding Irwin. The survey consisted of items relating to residents' identification with community, community attachment, community involvement and participation in the local economy. Using a randomized procedure, an adult member of the household was selected to complete the questionnaire. An overall response rate of 60% was achieved; response rates for town residents (77%) were higher than those for open-country residents (53%).

Physical Characteristics of the Community

Irwin is picturesque. The village is laid out in a gridwork pattern with street names such as Ellen, Ann, Ada and Eva, the names of the daughters of the village's founder (E.W. Irwin). At the northernmost edge of the community is the educational complex. This is the site of Irwin's original school building, which currently serves as Irwin's primary school. Also in this area is a newer building, originally home to Irwin's junior and senior high school, which currently houses the middle school of the newly consolidated school district, and assorted athletic and intramural fields. Toward the southern edge of the village are two grain elevators and a farm supply that are the anchors of the community's economic base, and a residential area consisting of house trailers and assorted older homes in various stages of disrepair. The southern edge of the community is also the location of a small park and the high school softball diamond, bounded on its perimeter by the east branch of the Nishnabotna River.

Entering the eastern edge of the community, one is greeted by a freshly painted sign that optimistically proclaims that 'Irwin is Wide Awake and Growing!' a claim that is at least partially validated by the immediate presence of Robinson's, a large, successful and newly remodelled implement dealer that serves a regional clientele in Deutz–Allis sales and service. Travelling into Irwin from the west, the visitor first sees the typical small town Iowa convenience store, and then quickly confronts the intersection of the village's business district. The physical dimensions of Irwin are about six to seven city blocks from east to west, and about eight to ten city blocks from north to south (Fig. 4.2).

The business district covers the better part of two blocks and is the location of much of the community's economic activity. On a normal day,

Fig. 4.2. Irwin, Iowa, in 1994. This image shows Irwin's business district in 1994. While business activity in the town has declined sharply, the physical structure of the town remains much as it was in the 1940s.

several of the diagonal parking spaces are occupied by pick-up trucks and cars as town residents and people from the surrounding countryside carry on business with the bank, the insurance agency, the post office, the appliance repair shop, the restaurant and assorted other commercial outlets. A number of farmers from Irwin and the surrounding area may be seen on the main street, interacting with one another over a cup of coffee in the grain elevator, farm supply or local restaurant, and carrying on business with local vendors. Several older residents lament the comparative lack of social activity along the main street as they remember Irwin when the community was a greater focus of residents' lives. The three or four empty buildings typically found in this area serve as testament to Irwin's steadily shrinking economic base and reflect the fact that the community is evolving from a full service community to one more specialized in meeting the changing needs of the surrounding agricultural hinterland.

With the exception of the extreme southern edge of town, the residential areas of Irwin offer a pleasant and diversified neighbourhood mix. The tree-lined streets that lead northward from the business district are attractive and contain well-maintained wood-framed homes, many of which pre-date the original study. Newer ranch-style and split-level homes are more typical of both the western and eastern edges of the community. Many Irwin residents speak of these clean, safe and attractive neighbourhoods as being among the community's most important assets and perhaps a key to the long-term viability of the community.

Agriculture

The west-central part of Iowa is made up of gently to moderately rolling hills with a highly fertile black loam soil composed of finely ground rock deposited during the glacial age. The area is also blessed by plentiful rain, averaging an annual 32 inches. For the first half of the 20th century, the area supported a highly diversified agriculture, consisting of farms that were generally a quarter section (160 acres) in size, with members of the family unit providing the bulk of management and labour inputs. Farm dimensions clearly reflected the Homestead Act legislation and were very typical of midwestern farms of this era. At the same time, especially during the era's latter stages, mechanization was starting to bring about structural changes that would have a profound and escalating impact on the social organization of farming and rural communities. Even as early as 1940, the effects of mechanization could be felt as machinery (particularly tractors, which were owned by slightly more than 60% of Shelby County farmers) made it possible for farmers to consider increasing the number of acres they farmed (Fig. 4.3).

Another significant change that characterized 1940s agriculture in this part of the state related to the changing patterns of land ownership. The authors of the 1940 study noted that one of the 'tragic' challenges to the traditional concepts of people in this community involved the sizeable increase in farm tenancy that had occurred since the settlement of the area some 60 years earlier. Then, during these early years of Irwin's settlement, nearly 80% of all farms were owner-operated; by the time of the study in 1940, slightly over 50% of the nearly 2150 farms in Shelby County were operated by tenants (Moe and Taylor, 1942). The authors of the original study spoke of the declining attachment of farm families to the land, a condition that they argued led to an increasingly tenuous relationship with the community.

Farms in Shelby County were highly diversified crop and livestock operations in 1940. Crops on the typical farm were in a four-field rotation, with two fields planted to maize, one to oats and one to forage crops such as lucerne or clover. Moe and Taylor indicated that crops were evaluated by the extent to which they supported the livestock enterprises, a process perceived as adding considerable value to basic crop commodities. Farms varied in the number of livestock enterprises, but typically included a combination of hogs, beef cattle, dairy cows, sheep and poultry. Hogs were considered the best 'money maker,' but most farmers preferred cattle. It is interesting to note that the typical Irwin area farm in 1940 was organized as a largely self-sufficient system, with the majority of crops being fed to the livestock, and crop rotations and animal waste the primary sources of nutrients for crop production. Contributing to the farm family's self-sufficiency was the near universal presence of a large garden that produced the staples used throughout the year.

Fig. 4.3. Irwin, Iowa, in 1941. Farming in the early 1940s reflected both traditional practices and new technologies. Here, a field is being worked by both horse-drawn equipment and a tractor.

Agriculture in the USA underwent a fundamental transition in the 20th century; these changes have had a profound impact on many of the nation's small and rural communities. Labour-saving technologies have generally been cited as the most important source of such changes, with the primary vectors being a dramatic decline in the number of farms and a corresponding increase in the average size of remaining farms.

Such a transition is uniformly descriptive of Iowa agriculture. From 1900 to the latest Census of Agriculture, Iowa farm numbers have declined by more than 60%, from roughly 228,000 to 90,000 farms, while the average size of operation has increased sharply, from 151 to 343 acres. At the time of the original study, 16% of the state's farms were under 50 acres in size; another 58%, from 50 to 179 acres; 25%, from 180 to 499 acres; and only 1%, over 500 acres. According to recent Census statistics, the percentage of farms under 50 acres had increased slightly to 18%, but the most dramatic increases over this time period occurred in the 500–999 acre category (625% increase) and the 1000+ acre category (3800% increase). These increases came at the expense of small to medium-sized family farms, as indicated by the fact that over half of the state's farms fell into the 50–179-acre category in 1940 while only 27% were in the same category in 1997 (Hanson *et al.*, 1999).

Iowa farms have generally become less diversified over time. The 1997 Census of Agriculture indicated the predominance of maize and soybeans, with almost 70% of Iowa farms harvesting maize and nearly 66% harvesting soybeans during that year. The decline in the number of diversified grain–livestock farms is indicated by the fact that in 1997 only a little over 40% of Iowa farms sold cattle, only 20% sold hogs and pigs, just under 5% had dairy cows and only 2% had chickens.

Reflecting the fertility of its farmland and the large percentage of its farms growing maize and soybeans, it is not surprising that Iowa continues to rank first in the nation in the production of both commodities. However, there has been an increased concentration of the state's livestock industry, as indicated by the fact that Iowa continues to lead the nation in hog production, and ranks second in egg production, sixth in cattle production, and twelfth in the production of milk, despite the rapid decline in the number of farms engaged in these enterprises.

Changes in Shelby County's agriculture generally mirror those for the state as a whole. Farm numbers declined from 2150 in 1940 to 921 in 1997, while during the same time average acreage increased from 173 to 372. Family farms account for 87% of the total cropland operated, significantly higher than the statewide figure of 75%. The relationship of Shelby County farmers to the land has also seen substantial changes over the past half century as the number of tenants fell from 50% in 1940 to 21% in 1997. Recent figures indicate that roughly one-third of the farm operators in Shelby County own all the land they farm, while a little over two-fifths are classified as part owners. The percentage of part owners has grown substantially since the farm economic crisis of the mid-1980s, increasing from 29% of Shelby County farmers in 1978 to over 42% in 1997. The majority of rented acres are farmed on a cash rent basis, reflecting the recent volatility of grain prices.

Eight out of ten Shelby County farmers harvested both maize and soybeans on their farming operations in 1997. The historical popularity of cattle is still evident in this part of the state, as 46% of the area's farmers maintain cow/calf operations. Other livestock enterprises have seen dramatic declines – hogs are present on less than 25% of the county's farms and fewer than 2% of area farms are involved in either dairy or poultry farming.

At the time of this restudy, Shelby County had not experienced the invasion of large-scale hog confinement units so evident in other areas of the state. Indeed, not only are fewer Irwin-area farmers raising hogs, but the total numbers of hogs sold have also declined. From 1992 to 1997, the number of hogs marketed in Shelby County declined from in excess of 400,000 animals to about 260,000. It is quite clear that larger operations now account for an increasing proportion of the county's total production – over 75% of the area's hog and pig production occurs on farms selling at least 1000 hogs annually.

It is instructive to compare these figures with Hamilton County (located in north-central Iowa), which has experienced the largest growth in large-scale hog confinement operations in the state. Between 1992 and 1997, Hamilton suffered a significant decline in the number of farms that raised hogs (from 235 to 159 farms). Unlike Shelby County, however, the number of hogs marketed during this same period grew by 385,000, to over 880,000. Further, while 76% of Shelby County's hogs were raised on operations selling more than 1000 animals, the comparable figure for Hamilton County was over 96%! Many Irwin-area farmers are convinced that it is only a matter of time before similar structural changes occur in their community. If the pattern characteristic of the north-central part of the state makes its way to western Iowa, it will be difficult to anticipate all of the changes that will be visited upon the region's rural communities (Fig. 4.4).

Irwin: Continuity and Change

When asked about changes in the relationship of area farmers to the Irwin community, a farmer who lived nearly 2 miles south of town said, 'it's changed in the respect that there are very few farmers left'. Looking out over the rural landscape, he reminisced:

> If you go over on this road, well, just looking up this valley here, I mean, Walt Hansons lived there and Ted Axtons over there and Neil Nelsons there, and, of course, Bill's place up there – there were two places up there across the road and one up here on the hill. I mean there is nobody there anymore! Buildings and all are gone in some cases. If you count the number of farmsteads along this one four-mile stretch of road, there were 14 back in 1940 and now there is only one.

Fig. 4.4. Irwin, Iowa, in 1994. Farming remains important in this part of Iowa, but the scale of agricultural operations has grown while the number of farms has declined.

Driving down this country road, the systematic depopulation of the countryside has changed the rural landscape forever, as symbolized by a rusting Farm Bureau sign and a lonely, creaking windmill. Such remnants of a once-vibrant Iowa farmstead now offer only a hint of the rich social texture of Irwin's former farming community.

Fewer farms, fewer farm families and smaller families – it is easy to idealize the nostalgic images of community emanating from such observations of change, but each of these factors, endemic to much of rural America, comes at a significant cost to the objective health and vitality of the rural community. The open-country population of the four townships immediately surrounding Irwin has declined by over 1700 (from roughly 2700 in 1940 to slightly over 1000 as of the latest census count). Another Irwin farmer lent meaning to these figures when he stated: 'All those kids and all those people [from the countryside] went to school, went to the churches, bought groceries and bought tires, bought gas, paid electricity bills and generally had a big impact on Irwin.'

Despite the changes in the agricultural hinterland, Irwin's current population is nearly identical to that registered in 1940. The 1998 population estimate for the village of Irwin was 379, only slightly lower than the official count of 381 in 1940. Irwin's population peaked at 446 in 1970 but has decreased in every census count since then.

It is important to recognize that 52% of Iowa's 953 incorporated communities, including Irwin, contain 500 or fewer residents, and account for roughly 5% of the state's population. Four in five of these communities (80%) suffered population losses between 1980 and 1990. Irwin, like many of these small communities, was settled as a service centre, meeting the needs of the agricultural hinterland. Its numbers of schools, churches, businesses and governmental units, and the people needed to run them, expanded as agriculture became increasingly sophisticated and integrated into a national and international network. The long-term sustainability of such service communities has been threatened by the precipitous decline in the number of Iowa farms. That Irwin's population peaked as late as 1970 indicates that institutional and population adaptations to the downward cycle have lagged behind systematic depopulation in the agricultural hinterland over the past six decades. We return to this point later as we discuss institutional change.

Over the past few decades, Iowa's population has become increasingly concentrated – its two largest metropolitan areas account for 43% of its population growth since 1990. At the same time, 72 of the 99 counties experienced at least some population loss between 1970 and 1998 (*Des Moines Register*, 13 March, 2000). What has been the impact of these changes for rural communities? One observation is that they have helped fuel a subtle but increasingly visible conflict between the state's rural and urban areas as debate has emerged over the distribution of limited state resources to education, recreation and highway construction projects. In this debate, some have advocated a triage approach that would make heavy investments in those areas most likely to contribute to economic and population growth. Others have argued that favouring urban areas is likely to continue the trend toward population concentration, contribute to a growing malaise in the countryside, and cause stagnation in the rural hinterland. Thus, Iowa's highly populated urban areas would thrive, often at the direct expense of the state's more isolated rural communities (*Des Moines Register*, 16 March, 2000).

Kaufman (1959) reminds us of the importance of the demographic context as it influences the nature of community identification and participation. Demographically, Irwin looks like many other rural communities in this region. Its age and gender distribution closely resembles an inverted pyramid, with almost 25% of the community's population aged 65 years and older, and only 13% under 10 years of age. The most noticeable narrowing is in the young adult age category (20–24 years) that accounts for only 3% of Irwin's population. The gender distribution is roughly equivalent for all age groups, except for the 75 and older category where the number of women is almost double the number of men. Racially, Irwin is exceedingly homogeneous, with only a smattering of minorities represented in its population.

One conclusion reached by Moe and Taylor was that Irwinites placed a great deal of importance on educational attainment. Generally, this emphasis on educational achievement is still in evidence in the conversations of local residents, yet only 12% of adults aged 18 and over report having earned a bachelor's degree. Another 20% have attended some college or have received an associate's degree from a community college. This pattern is more a reflection of the 'drain' of the community's youngest and best-educated residents, who leave the local area in search of enhanced career opportunities in the state's and nation's larger cities, than an indication of a decreased emphasis on education. Interestingly, this pattern was also identified in the original study, where statistics reported that nearly 60% of the high school graduates in the 10 years preceding the study moved from the community, despite a general wish to stay (Moe and Taylor, 1942, p. 70).

About 66% of Irwin's adult residents are employed for at least part of the year. Almost 75% of the adult males are employed. Fewer than 60% of the adult females are employed, although the number of women in the workforce has been growing noticeably in recent years, as rural women come increasingly to resemble their urban counterparts. Many of the jobs, particularly those held by women, are lower-paying service jobs where, according to one young mother, the 'wages are scarcely enough to pay for the commuting and child care expenses. The only way I can justify working away from my home is because of the benefits, particularly the health insurance'. The large majority of non-working Irwinites are in their retirement years, with very few actually unemployed. The largest general occupational group is the professional service worker, accounting for 25% of the employed residents. This observation is buttressed by the fact that the local school system is the community's single largest employer. Retail trade, agriculture, and business and financial services are other significant occupational categories.

A very large majority of Irwinites live and work in Shelby County, but more than 40% of the employed adults work outside of Irwin – many in the county seat town of Harlan. Irwin's median household income as of the latest census was US$23,250, roughly equivalent to Shelby County as a whole, but some US$8000 lower than the average yearly incomes reported by residents of Iowa's most urban counties. Roughly one in ten Irwin residents lives below the official poverty line, with a nearly equivalent number receiving some form of public assistance (notably children and senior citizens). Those few persons of working age who are on public assistance programmes, often single parents from neighbouring cities who are on AFDC (aid to families with dependant children) and seeking the cheap housing offered by the region's smaller communities, are accepted but not well integrated into the local community, either by personal choice or circumstance. This occurs as the most noticeable and rigid feature of a social class system which, despite objective differences in lifestyles, life chances and associations, is downplayed by local residents and not a highly recognized component of the community's social structure.

This demographic profile and description of changes in the area's agricultural structure provide the contextual backdrop for the following discussion of Irwin as a dynamic interactional field and institutional complex. It will become obvious that these demographic and ecological changes have placed limitations and otherwise influenced the number, variety and quality of fields that make up Irwin's contemporary social structure.

The Social Dimension of Contemporary Irwin

In what has become a classic definition, Roland Warren defines community as 'that combination of social units and systems that perform the major social functions having locality relevance' (Warren, 1978, p. 9). In this social systems approach to the study of community, Warren concentrated on several key functions – production and distribution of local goods and services, socialization or the transmission of knowledge, social control, social participation and mutual support. Warren understood that these and other functions, while important to the locality, were not functions over which the local community had complete control. He suggested that communities were organized 'horizontally' in the interest of providing a local network response to meeting the basic functions, but, to one degree or another, communities were also dependent upon a vast and increasingly complex 'vertical' system that tied the local community to the external world to help in the process of providing for these locality relevant functions (Warren, 1978).

Kenneth P. Wilkinson (1986, 1991) concentrated on the 'organization of local life' and the identification and analysis of a 'field' of community actions that converged to meet recurrent and spontaneous community needs. Rather than focusing on institutional patterns *per se*, Wilkinson's approach to the study of community targeted the interactional patterns of local community residents and groups as they collectively sought to deal with emergent community issues. Wilkinson stressed that local community members were simultaneously connected to multiple special interest fields (in some cases external to the locality), but that this did not negate the 'tendency for people who live together to interact with one another on place-relevant matters' (Wilkinson, 1991, p. 37).

Institutional structure

Steenhusen's Itco, a small, now-vacant hardware store located in an ageing metal building about one block west of Irwin's downtown district, was until recently a contemporary link to Irwin's past. Paulsen and Steenhusen's General Store opened its doors in 1888 and for the next 100 years survived by adapting its structure and function to the area's changing social and economic landscape. From the time of the original study until 1969, the patriarch of the Steenhusen family, Peter, and his son John H., built a thriving hardware and farm implement business that specialized in the sales and service of John Deere farm machinery. As John Deere adopted a regional marketing strategy, Steenhusen's and dozens of other Deere dealerships in Iowa's smaller towns closed their doors for good in the late 1960s and early 1970s, roughly the same period that farm machinery became larger and more efficient and the number of farmers began to decline sharply. The business continued to be operated as a hardware store, with ownership transferred to Peter Jr in 1967 and finally to John L. Steenhusen, both representing the third generation to be affiliated with the business. During the 1980s and early 1990s, the volume of business declined; they finally closed operations in 1995.

In 1940, Steenhusen's was one component of a relatively complete institutional structure that provided for the production and distribution of goods and services in Irwin. As a service centre for the immediate agricultural hinterland, Irwin sported a complete array of farm-based businesses and services, including animal feed and seed operations, a creamery and produce business, a grain elevator and a veterinarian. Complementing the farm businesses were two cafés, a general store, two grocery stores, a plumbing and heating business, a drug store, a furniture store, a post office, a telephone exchange, a hotel, a lumber yard, a bank and two railroads. For the recreation and entertainment needs of the local population, Irwin had a movie house, a pool hall, a tavern, and an opera house where travelling troupes performed plays and community-wide dances were held.

Then, the community was also the location of the Irwin–Kirkman consolidated school and three protestant churches. It was generally conceded that Irwinites could, with a few exceptions, get most of their daily needs satisfied within the physical confines of the village. Yet, even in 1940, Moe and Taylor lamented the loss of several local services and businesses. They concluded that, 'In general, the goods and services available in the community, at present, are those which are unspecialized, of low cost, and frequently demanded' (Moe and Taylor, 1942, p. 64).

Changes in Irwin's institutional network have been rather dramatic over the past five decades, generally continuing the trends identified by Moe and Taylor in the 1940s. At its most basic level, Warren (1978, p. 171) argued that a community needed to be able to provide 'for the exchange of goods and services among units in the community' and provide opportunities to its members for productive work. Wilkinson (1991, p. 70) also spoke about the need to provide for 'sustenance needs' before members of the locality could turn their attention to satisfying higher-order social needs. According to Wilkinson, 'this demand [for sustenance needs] generalizes to needs for jobs, income, markets, homes, and a range of services'.

Irwin's business community has, on the whole, become more specialized, showing continued strength in the farm sales and service sector but struggling to maintain a local presence in most other areas of retail trade. Today, Main Street has a specialty meat locker, an insurance agency, a bank, a post office, a beauty shop, a restaurant/tavern, a store that sells and repairs small appliances, and a pool hall. A convenience store stocked with basic food staples and a service station round out the formal business community. During the 3 years that field data were collected, Irwin lost two restaurants, an insurance agency, a feed store and the services of a veterinarian. As in most small communities, Irwin boasts a somewhat active informal economy exemplified by a local resident who built a substantial reputation in the area for cake decorating and a retiree who builds wooden hope chests that are marketed widely.

In the agricultural sector, Irwin has two rather active grain elevators, a farm supply co-op, and a Deutz–Allis implement dealer that has recently added a line of hardware, helping to fill the local void caused by the closing of the Itco store. Most examples of successful local businesses, including the implement dealer, have opted for a regional marketing and service strategy to remain competitive. Robinson's, the regional Deutz–Allis dealership, serves an exclusive area bounded by Minnesota and Missouri to the north and south, beyond the Nebraska border to the west, and nearly to Des Moines to the east.

There has been much discussion about the damaging effect of regional retail centres on business activity in the nation's small and mid-sized communities. Iowa State demographers (Stone et al., 1999, pp. 301–314) have calculated an estimate of the net gain (loss) in retail sales measured at the county level. In 1998, 16 of Iowa's 99 counties had higher retail sales than would be expected from state averages, meaning that they gained sales from other counties. These mostly urban counties captured between US$50 million and US$2 billion in sales from residents outside their borders. Shelby County, on the other hand, lost an estimated US$40 million in retail sales that same year, compared to a surplus of US$190,000 in retail sales that was registered in 1980. Small communities like Irwin have suffered the loss of significant business activity to county seat towns, including Harlan in Shelby County, but these figures indicate that many county seat towns, in turn, have also suffered losses as county residents leapfrog local alternatives in search of more variety and lower prices.

In our survey of Irwin area residents, a certain measure of loyalty to local business services was evident, but it was also clear from many respondents' comments that this loyalty was not complete. A farmer who had grown up in the area and had attended school in Irwin stated that 'I guess I'm pretty loyal to Irwin, but there are so many things that are not available that a person is forced to go elsewhere.' This statement is corroborated by the survey data which show that upwards of 75% of Irwin area residents travelled outside the local community for such basic services as groceries, health care and recreation. Even for goods and services that are more generally available, such as banking and farm supplies, fewer than 50% of the community's residents relied exclusively on local businesses.

Warren (1978, p. 189) listed social participation as one of the key locality-relevant functions. While arguing that the community must provide its members with opportunities for participation, Warren recognized that the range of alternatives available in communities would be highly variable at the local level. Because churches are examples of voluntary associations that typically have the greatest number of local members, and because they relate to an important dimension of the community member's social life, Warren used this institution to exemplify the horizontal and vertical dimensions of social participation.

Churches were among the first social institutions to be established in Irwin. Moe and Taylor reported that residents, even those who did not regularly attend services, considered churches to be the most important institutions in the community as they contributed in a unique way to the residents' quality of life (Moe and Taylor, 1942, p. 60).

There has been a relative stability in the number of churches since 1940, despite the rather sizeable loss in the area's population. The Lutheran church has the largest membership, followed by the Methodists and the Church of Christ. However, less than 66% of Irwin's residents maintain membership in one of the local congregations. Part of this leakage is related to denominational affiliation. Several neighbouring communities, for example, have a Catholic church, which Irwin lacks, while other residents have long-standing family affiliations with churches in some other community. Irwin's Lutheran church and Church of Christ both have full-time pastors, although the Church of Christ has had difficulty in attracting and keeping a pastor due to its dwindling membership. Irwin's Methodist church has adopted a regional strategy of survival by sharing its pastor with two other congregations in nearby communities. An example of Warren's notion of 'horizontal integration' which relates to strategies of cooperation and adaptation is seen in the development of an ecumenical approach, in which some religious and secular events are jointly celebrated across congregational lines. As was the case in 1940, Irwin's churches continue to play an important role in the social and spiritual lives of its residents, although demographic changes are making it increasingly difficult to provide for a full range of alternatives.

Given a choice, most residents of Iowa's smaller communities would readily list the local school as the most important link to their community's future. There are dozens of stories of the emergence of nasty conflicts in Iowa's rural communities when the future of the local school was threatened by the spectre of school consolidation. The reason for the intensity of these feelings is linked to the multifaceted role that the school plays in the community. Warren (1978) linked several of his locality-relevant functions to the school, including socialization (the transmission of society's knowledge, values and behaviour patterns to succeeding generations), social participation and social control. As the literature has demonstrated, the school also exists as a primary focal point for the symbolic attachment of residents to their community. From Wilkinson's (1991) field perspective, schools, almost by definition, are locality oriented. Nor is it surprising that issues relating to schools, such as local control, have become the focal point for significant community interaction and the arena for social action at the local level (Fig. 4.5).

Fig. 4.5. Irwin, Iowa, in 1941. One-room school houses like the one pictured here were the norm at the time of the original study of Irwin and were scattered throughout the rural countryside.

Recent dramatic change has occurred in Irwin's school system. When this restudy effort began in the early 1990s, the Irwin–Kirkman consolidated school district had a comprehensive K-12 educational system, with the elementary school housed in the original consolidated school building and the junior and senior high schools located in a modern facility directly to the north of the original building. Class sharing, plus the sharing of some minor sports and other school activities, were already under way, which prompted local residents to contemplate the future of Irwin without its own high school. Several neighbouring communities had lost schools through consolidation and many Irwin residents argued that such an action would signal the death knell for the local community. One resident complained that the loss of the high school would result in the loss of 'the focal point in the community which means that town and country people will begin to cut their ties to Irwin'. Another spoke of the 'devastating impact on property values' that would follow the loss of the high school.

The road to consolidation with Manilla, a town roughly twice the size of Irwin and located about 12 miles to the north in Crawford County, began slowly with the sharing of a play, then an art teacher, and then junior high athletic teams. In the late 1980s, whole grade sharing began. In the latter stages of the whole grade-sharing plan, it was determined that the newly constituted high school would be located in Manilla the first year, in Irwin the second year, and again in Manilla for the final 2 years before the planned formal consolidation. One of the incentives for sharing was that the districts were offered over US$1 million in direct aid from the Iowa Department of Education to carry out the cooperative agreement (Fig. 4.6).

Fig. 4.6. Irwin, Iowa, in 1994. This building originally housed all grades, but is now limited to the primary grades. The consolidation of school systems has resulted in the closure of many smaller schools and has been a focus of substantial community controversy.

Both internal and external pressures for the broader consolidation with Manilla became evident in the unfolding of this story. According to Irwin's superintendent, low student numbers in the Irwin–Kirkman district prevented it from offering the broad and competitive curriculum demanded by residents for their children; moreover, the system had increased difficulty keeping up with state-imposed curriculum standards due to its lack of resources (also associated with small student numbers).

The two school boards employed a panel of school reorganization experts to conduct a neutral assessment of the potential for the consolidation and to make site recommendations for the permanent location of the high school. After a comprehensive study of comparative advantage, the panel recommended that each community maintain its own elementary school, that Irwin become the permanent site of the new district's middle school, and that Manilla become the location of the high school. Further, they recommended that all home football and girl's softball games be played in Irwin and that Manilla be designated as the site for home basketball games and other high-school sporting events. When the plan went to voters, it received 96% approval in Irwin and 97% approval in Manilla. Kirkman, located about 8 miles to the south of Irwin, rejected the proposal, with fewer than 50% voting for the plan.

On 1 July, 1992, the two school systems officially merged. In a local newspaper account (*The Daily Nonpareil*, 29 January, 1992), several students and teachers reflected on the merger. One teacher stated that at the beginning of the sharing agreement, there was an awareness of 'which students were from Irwin and which were from Manilla. But now they're just kids'. A student recalled: 'at first, there were big rivalries among students, but that was over when you got to know everybody or got to be friends'. She also indicated that any remaining problems or stereotypes stemmed from older people in the community. By and large, the consolidation has gone smoothly, primarily due to the forward planning and complete openness employed by community leaders. As evidence of the favourable impact, the superintendent pointed to the fact that the sharing agreement has allowed the high school to begin offering vocational agriculture and a business applications course, and that art and Spanish were added to the elementary curriculum. There has also been a dramatic improvement in the competitiveness of some of the athletic teams as a result of enhanced bonding between the towns in the district. The school spirit stands in stark contrast to the comments of an Irwin resident who, before the consolidation, reminisced about the long-standing rivalry between Irwin and Manilla when he said that 'I won't be able to stand seeing my kid dressed in green and white!'

There is an understandable reluctance on the part of rural community residents to quickly embrace school reorganization plans. As Warren (1978) reminded us, the school is the primary formal socialization agent at the local level and has made a significant impact on the development of a young person's world view. Along with the church, the school plays an important role in social control, particularly among the community's younger generation. In some people's eyes, to relinquish these functions to an 'external' agent was to lessen the influence that the locality could exercise on the socialization of the younger generation, and perhaps even result in a loss of control over the future state of the community. Here, however, the consolidation with Manilla was seen by most residents as inevitable and decidedly preferable to the alternative of merging with Harlan to create a countywide school district.

Wilkinson (1991, pp. 85–86) discussed the idea that rural dwellers were more likely to have larger numbers of 'strong ties' or durable and permanent relationships, as exemplified by strong friendship and kinship linkages (Granovetter, 1973). Wilkinson suggested that the reason for this preponderance of strong ties was not merely a matter of preference as much as a reflection of rural settlement patterns that tended to restrict the number of strangers (weak ties) a person in a rural location had an opportunity to meet and interact with. As the urbanization process continued, however, the potential number of 'weak ties' increased, creating the possibility for new networks and opportunities as local control became increasingly threatened.

One theme that emerges from observations is that the meaning of the term 'local' appears to have changed as a result of the evolving regionalization of several community services, including education. One interpretation of Wilkinson's work is that school consolidation would broaden the interactional field in rural communities beyond traditional geographical boundaries. This was apparently the case in Irwin, where several residents indicated that, as a direct result of the consolidation, they were interacting more regularly with Manilla residents and more frequently using the services provided by their neighbouring community.

This is not to say that the transition is complete and that Irwinites have uniformly forged this broader identity. Some residents seemed concerned that school consolidation would put pressure on primary ties and contribute to the erosion of local community autonomy in carrying out the functions of socialization and social control. A larger number, however, seemed convinced of the potential for building a stronger system of resources through consolidation, and of creating for their children the kind of bridges to mobility made possible by immersion in a broader social system composed of both strong and weak ties. For this group, community is beginning to assume a broader meaning and the new definition is emerging as the legitimate arena for local participation and social action.

Irwin as a local social system

It is quite clear from their responses to the community survey, as well as more informal comments, that most people live in Irwin by choice and not because they are trapped in the community. Indeed, several indicated that it might be more convenient for them to live elsewhere, for job-related or medical reasons, but that they none the less planned to stay in Irwin.

The question of attachment to community has been a popular focus of community studies for several decades (Kasarda and Janowitz, 1974; Goudy, 1990; Beggs et al., 1996). A related concept, psychological identification, was one of the four dimensions chosen by Warrren (1978) to differentiate modern American communities analytically. Moe and Taylor spoke of the strong bond that Irwinites had with their community in the 1940s. They included statements from residents in all age groups about passionate attachments to Irwin. These testimonials spoke of the friendly, honest, upright character of their neighbours. It is not surprising, then, that only a lack of opportunity would cause them to consider leaving Irwin (Moe and Taylor, 1942, p. 72).

In the restudy, questions about community attachment were part of the survey. Ninety per cent of the participants indicated that they felt 'somewhat or very much at home' in the Irwin community. There was nearly a 10% difference between town and country residents in their responses to this question, with town residents expressing more attachment. The two dominant responses (accounting for 40%) given as reasons for these feelings of belonging are that residents felt that their roots were in Irwin, and that the community was perceived as having very friendly and hospitable people. These were very similar to the sentiments expressed by residents in the 1940s. Nearly 85% of the survey participants stated that they would be either 'very or somewhat sorry' if circumstances forced them to move from Irwin. Again, a larger percentage (11% higher) of town residents expressed these sentiments than those living in the open country.

Even though Irwinites scatter to get many of their basic needs fulfilled or to pursue their careers, the base of their social associations remained closely tied to the local community. The average number of close personal friends living in Irwin for the total sample was nearly 30, but there were substantial differences between town (40) and country (24) residents. Two-thirds of the respondents claimed that half or more of their close friends lived in Irwin, but again there were large differences between town (85%) and country (56%). This difference also held true for the proportion of community residents known, with 94% of town residents and 71% of the country residents indicating that they knew at least half of the adults in the community.

When we asked about relatives, we discovered that both town and country residents averaged about 11 close relatives in the Irwin area. About 25% of both groups stated that at least 50% of their relatives lived in Irwin, but an equal percentage reported that none of their relatives lived in the area. Generally, people focused on the positive value of having relatives in close social and geographical proximity, but some commented on the downside of this high density of relationships. A younger town resident commented:

> It seems as if everyone is related in one way or another. That's okay, except, for example, if a manager of a business is managing his brother-in-law and the brother-in-law ups and quits or is fired, then it has ramifications for not only the business, but also the family and even the community.

Community attachment heavily influences residents' participation levels in local social action networks. Warren (1978) included general social participation as one of the five locality relevant functions of the community; Wilkinson (1991), while recognizing that the community field is only one interactional field in a local population, had a special interest in interactions of people around issues related to the locality. Responses to a series of survey questions on community involvement indicated that slightly over 50% of the respondents reported that at least one household member had participated in a local community improvement activity during the previous year. Town dwellers, on the whole, were considerably more active in local affairs than were open-country residents. This same town–country distinction was observed for membership in local organizations, with town residents belonging to an average of three organizations and country dwellers belonging to two. When asked about their level of local community involvement, a majority of the total sample characterized themselves as either 'somewhat' or 'very' active. Once again, town residents (62%) were more likely to describe themselves as active than were open-country residents (48%).

It is evident from comments made by many Irwin residents that there was a strong affinity with the local social system, although it was by no means complete. Even at the time of the original study in 1940, many examples of Warren's (1978) vertical ties were present, especially those linking the local community to external social and economic systems. With increased mobility, made possible by sweeping transportation and communication advances, it is clear that the strength of these linkages has not subsided. These pressures have been reinforced by the increased trend toward regionalization of many community services. In this process, the meaning of 'local' has become transformed. Despite this, a relatively strong bond linking Irwin residents to their local community remains. Indeed, it is interesting to speculate that the relatively high levels of local sentiment may reflect an enduring *gemeinschaft* spirit that persists in the wake of an increasingly *gesellschaft* society. Also, this sentiment appears to have a link to the observed tendency for Irwin residents, particularly town dwellers, to participate actively in what Wilkinson refers to as the community field.

The differences in social interaction and community involvement between town and country dwellers is interesting and worth further examination. Moe and Taylor concluded from their study that the community of Irwin was more of a community in 1940 than it was when the village was established 50 years earlier. They pointed to examples of the development of stronger relationships between town and country residents as evidence of the stronger community bond. The authors concluded that instances of farm kids being discriminated against in the local school, property disputes and feelings that local merchants were unfairly profiting from the labours of farmers were earlier points of contention that were slowly being supplanted by increasing cooperation and mutual involvement in the community's institutions and feelings of mutual dependence between town and country dwellers.

Yet, the survey findings for the present study indicate that, in nearly every measure of social participation and attachment, town residents consistently reported higher levels of engagement. One reason for this pattern is that the open-country residents lived in a relatively large geographic area and in some cases lived closer to other communities. For example, a majority of the residents of Kirkman voted against the school consolidation measure with Manilla, since they felt a closer affinity to Harlan, the county seat. In fact, because of the issue of proximity, several Kirkman area residents enrolled their children in the Harlan district after the consolidation effort.

Another potential explanation for the increased strain in the relationships between open-country residents and the local community is rooted in local farmers' feelings about property taxes and school funding. Shelby County farmers are an ageing population. Given that many of these farmers are now in the latter stages of the family life cycle, they do not have as direct a link to school activities and may be actively opposed to increasing

Fig. 4.7. Irwin, Iowa, in 1994. This modern-day grain elevator operation located in Irwin reflects the continuing importance of agriculture to the local economy.

property taxes (that they already perceive as too high) to improve school facilities and curriculum.

In addition, zoning, the siting of livestock facilities and other land-use issues have often pitted the interests of farmers against those of town folk, and may have contributed to a further erosion of open-country residents' relations with the local community. And, farmers' trade patterns have undergone a continuous change. The trend to larger farm operations often means that local farmers' needs for inputs and services may exceed the capacity of the community to respond. As a result, farmers no longer relate to a single community. They may go to one community for banking and financial needs, another to purchase livestock feed, another to market their crops, another to attend church and yet another for recreation and entertainment. The comparative ease of mobility, added to rational business decisions, have lessened the high degree of mutual dependency between town and country that Moe and Taylor documented, and may be an important factor in explaining some of the observed differences in community attachment and participation. When analysed from an institutional perspective, this is an additional example of functional community boundaries being expanded to fit a broader regional perspective, and highlights the potential impact that these changes have on identification with the local community.

The community field

Irwin residents point with pride to several visible examples of widespread local participation that have contributed to a special kind of community solidarity. Centennial celebrations often have this effect and Irwin offers a case in point. The summer of 1981 was the official observance of Irwin's centennial, although various community groups had started planning months before. On Friday 5 June, a pageant entitled 'Pages of Time' was held. It had nearly 200 residents participating in skits that centred on a significant historical event in each decade of the town's existence. On Saturday, there was a big parade with 325 entries, many from surrounding towns that were also celebrating their own centennials, followed by a dance that evening in the new community building. On Sunday, community church services were held.

The new community building and volunteer fire station were intended to be built from the proceeds of the centennial, but Irwin residents decided that they wanted the building completed in time for the celebration. When this decision was made, volunteers fanned out across the countryside and through the town asking for US$500 pledges. In 2 days, they had collected pledges totalling US$60,000. Ground breaking for the facility was 5 March, 1981. The total cost for materials was US$105,000; through the combined volunteer labour of townspeople and farmers, the community building was finished by the time of the centennial.

The Cozy Corner, a café on the north-east corner of Main Street, offered local residents a place to have a morning cup of coffee and a rol!, to eat lunch or dinner, and to meet cronies to have a beer after a day of work. The young couple who owned it lived in the back of the café in a small apartment. About 8 a.m. on a Monday morning in July 1995, a person going to work noticed smoke coming from the front of the building. Within minutes, flames engulfed the entire structure and volunteer firefighters from Irwin and several surrounding communities gathered to fight the blaze. Despite the efforts of the firefighters, the building was a total loss, but thankfully no one was injured. A complicating factor was that Irwin's museum, housed in the old hotel building and filled with fascinating local artefacts, was located immediately to the south of the café. A small army of residents removed everything from the museum that Monday and when the building was out of harm's way, moved everything back 2 days later.

The young couple lost everything in the fire. They secured a small business loan to rebuild the restaurant, but lacked the resources to rebuild their dwelling. Within the month, the chamber of commerce had mobilized the community. By sponsoring a fund-raising dinner, holding a silent auction, and accepting free-will offerings, they gathered more than US$16,000 to help the couple rebuild their apartment.

These examples of local community action offer just a few instances of local pride. They are not meant to idealize the rural community, but to offer examples of the successful mobilization of the community field to confront local issues. Irwin is not perfect. Indeed, during the years of the farm crisis, several people recounted a lack of local response to farmers who were in financial and social difficulty, and that farmers were often forced to go outside the local community for desperately needed support. Overall, however, community members seem to recognize that it will take a special effort and a large investment of social capital to play the survival game when the deck seems to be stacked against them.

Irwin's Future

Given the demographic and institutional trends and the social processes outlined in this chapter, what is Irwin's future? Will it continue to lose its population base and become more economically and socially marginalized? Will the community's young people continue to leave the area in search of economic and educational opportunities elsewhere? Will the community's institutional base continue on its path of increasing specialization and less completeness? As residents increasingly come to depend on other communities for satisfaction of their daily needs, will there be a corresponding loss of identification with the community? Or will social and economic forces be sufficient to sustain the community into the future (Fig. 4.8)? Even though future prosperity and growth for communities like

Fig. 4.8. Irwin, Iowa, in 1994. The increased concentration of commercial activity in larger population centres has resulted in the gradual decline and closure of many of the traditional businesses in small towns like Irwin.

Irwin seem only remotely likely, several examples of small communities successfully reversing their fortunes offer tantalizing evidence that the community's fate is not sealed. Based upon many conversations with community residents, it is obvious that while the community's logo, that 'Irwin is Wide Awake and Growing', may not be completely consistent with the evidence, there is enough optimism, community spirit, local participation and leadership to offer at least some hope for the future. Social attachments and interactions representative of community clearly persist in Irwin.

Much of the recent discussion among Iowa-based rural development specialists has centred on the potential of such things as e-commerce, entrepreneurism and value-added agricultural industries for contemporary rural development. They stress that given the significant changes in agriculture, farm-centred service communities face an uncertain future. When a group of Irwin's leaders came together to discuss future directions for their community, they mentioned several items, including school stability, dressing up Main Street, creating an annual celebration and cleaning up the cemetery as positive short-term measures. For the longer term, goals such as school stability, development of the housing stock, attracting a medical/dental clinic and industrial development dominated the discussion. In reality, there is a pronounced ambivalence about future development, as residents are caught between the feeling that the preservation of desirable community traits might be threatened by certain kinds of development and the fear that the community's fate might be forever sealed if significant changes in the social and economic base are not pursued.

Warren (1970) developed a checklist of local society qualities and argued that the imagery that residents held of their community could be an important reflection of community attachment and identity, as well as providing an indication of preferences for the future direction of development. Goudy (1983) used these descriptive items of the 'good community' in an empirical examination of perceptions of the ideal and real community. When Irwin residents were presented with these statements, most perceived that certain community characteristics (including townspeople knowing each other, participating in local community affairs, taking pride in their community and dealing effectively with problems) were descriptive of the ideal community as well as their actual home community. In other areas, however, there are significant disparities between the ideal and the real. For example, 75% of the respondents felt that community control over local affairs was characteristic of the ideal community, but fewer than 50% felt that this was descriptive of Irwin. Also, 75% of the residents felt that the absence of conflict was descriptive of the ideal community, but only about 50% felt that Irwin fits this description. In viewing equality issues, nine out of ten residents viewed the mixing of social classes and diffuse decision making as descriptive of the ideal community, but less than 66% viewed these conditions as characteristic of Irwin. Interestingly, only about 40% felt that similarity of residents was descriptive of the ideal community, but over 66% perceived population homogeneity as characteristic of Irwin.

It is quite clear that despite significant economic, demographic and institutional changes, Irwinites still exhibit relatively high attachment to their community and value the traditional primary ties that bind them to each other and to local society. It is also clear that local residents, while appreciating the historical and contemporary significance of agriculture for their community, seem to recognize that the relationship has been fundamentally and permanently altered. There is an emerging recognition that a positive future for Irwin will depend upon the ability of the community to seize new opportunities. In the process, they may need to sacrifice at least some autonomy over local decision making. Evidence of the functional and psychological expansion of community boundaries is quite strong. Existing transportation and communication systems are sufficient to allow for significant expansion in the definition of local society, but local residents must make a choice about their level of participation in this broader community and in the determination of what kind of future they want for Irwin.

Acknowledgements

I would like to express my deepest gratitude to the people of Irwin who gave generously of their time in helping me gain an understanding of the past, present and future of their community. Particular thanks go to Bill

Brue and Garland Barrett, without whose help this research would not have been possible. I would also like to thank my colleagues in Iowa State's Department of Sociology, particularly Willis Goudy and Paul Lasley, and my father, Otto G. Hoiberg, a community sociologist whose insight and experience provided inspiration and insight and to whom this chapter is dedicated.

Community Change in Harmony, Georgia, 1943–1993[1]

Gary Paul Green

The Setting

The objective of the Rural Life series was to 'investigate the cultural, community, and social psychological factors which either facilitate or offer resistance to change, contribute to adjustments and maladjustments, and to stability and instability in the individual and community life'. The Department of Agriculture selected Harmony, Georgia as one of the sites in the Rural Life series because it presented a strong biracial element. In the forward to the report, John Province indicates that 'it is, in truth, two communities, having little in common except the understanding that keeps them apart and their economic interdependence'.

Harmony is located in Putnam County, Georgia, which is one of the 35 counties in Georgia designated as the Old Plantation or Black Belt. Waller Wynne's analysis of the change taking place in Harmony in the early 1940s focused on the consequences of the break-up of the plantation system for the community. The plantation system, with its unique class structure and culture, had shaped the nature of community life in Harmony for almost a century. Under this production system, cotton was the primary crop, and almost all production was oriented toward the commercial market. Plantation agriculture was based on exploitive cultivating practices, and as a result, the land suffered, as did the people.

It was the boll-weevil that brought an end to the plantation system in Putnam County. The boll-weevil arrived in Putnam County in the early 1920s, leading to major changes in land use, the production system, the class structure and community institutions. Farmers were forced to diversify production and most tenants were pushed off the land. In 1919, Putnam

© CAB *International* 2002. *Persistence and Change in Rural Communities* (eds A.E. Luloff and R.S. Krannich)

County farmers planted 42,000 acres in cotton; by 1924, only 6000 acres were planted. Ten years after the arrival of the boll-weevil, the population of Putnam County had declined by 45%. Many land owners turned to dairying as their chief source of cash income. Wynne (1943, p. 18) reports 'in 1919 virtually the entire cash-farm income was from the sale of cotton; in 1924 almost half of the cash farm income was from the sale of butter fat, cream, and whole milk'.

This shift from cotton production to dairying reduced the demand for labour. Dairying, which had begun prior to Wynne's study, did not require as large a labour force as did the extensive cultivation of cotton. Few of the tenants moved into dairying, because of the high capital requirements and the regulations governing milk production. Most of the former tenants migrated to Atlanta or to northern cities such as St Louis, Detroit or Chicago.

Although Wynne examines the extent of economic and social change in Harmony that resulted from the demise of the plantation system, he was most impressed with the stability of the local community. John Province, in the forward to the report, indicated that Harmony was not the most unstable of the communities studied in the rural study series. Wynne (1943, p. 17) comments on the persistence of community even in light of the significant changes that had taken place in the region: 'Despite war, despite disastrous soil exhaustion and erosion, despite economic crises and depression, despite the infestation of the boll-weevil that made virtually impossible the profitable production of cotton and that brought the loss of nearly half its population – the Harmony community has endured.'

This view of the Harmony community is a bit curious because Wynne does admit that there were separate communities – one White and one African-American. His comments probably should be taken to mean that both the White and African-American communities endured despite the changes that had occurred in the region.

Wynne's account of the effects of plantation agriculture on the region also differs from those of others who had studied the social and economic systems of the Black Belt. For example, Arthur Raper (1936) studied Greene County, which borders Putnam County, just a few years earlier, and came to different conclusions about the fate of the Black Belt. He believed that the plantation system had left the region crippled, and predicted the emergence of a peasant rather than the growth of independent farmers.

Like Wynne, I am interested in how social relationships have changed as the community experienced structural changes over the past 50 years. In many ways the Harmony community continues to be influenced by the plantation system that shaped the development of this region. The plantation system left in its wake a region with few resources. Although the New Deal did much to provide a basic level of living for the residents in the area, it did little to change the structure of the community. Much of this region in the rural South experienced population and employment growth in the 1960s and 1970s. Communities in the Black Belt, however, were less fortunate in this regard. New industry did not locate in the region for several reasons, including the belief that African-Americans would be more likely to unionize and the lack of available services and amenities. In the 1980s, however, several manufacturing firms located in the region, which lifted the level of living among African-Americans.

One of the major influences on the Harmony community in recent years has been the development along Lake Oconee. The Oconee River runs along the eastern border of Putnam County and a dam is located just south-west of Harmony. The lake now covers much of the land that was originally part of the Harmony community. This dam, which was built in the 1970s, has spurred real estate development near Harmony, mostly in seasonal and recreational homes for people who live in Atlanta and nearby cities. Putnam County is located within an hour's drive from Atlanta, Athens, Augusta and Macon. The new residents of the Harmony community are often homeowners in these fairly expensive developments. Lake residents are much less dependent on the local community for their social and economic life. Seasonal residents may become very attached to their community through lakes associations and contacts with other seasonal residents (Green *et al.*, 1996). This interaction, however, does not produce a shared interest with the larger community. A result of these processes is a growing division in community life between lake residents and full-time residents.

To examine community life in Harmony in the 1990s I use several sources of data. First, I conducted interviews with key informants in Putnam County. These interviews were conducted with local business leaders, government officials and others knowledgeable about the community. The interviews were fairly unstructured but usually included the same set of questions regarding experiences with, and feelings about, the local community, major community issues/concerns/problems from the informant's viewpoint, and anticipated/preferred future changes/trends affecting community life that were used in the other communities in this study.

Secondly, I conducted a mailed survey among 500 Putnam County residents. These households were selected randomly from the voter registration lists in the county. Reminder letters and replacement questionnaires were mailed 2 and 4 weeks after the first mailing. This procedure produced a response rate of 59%. The questionnaires obtained information on the social and economic characteristics of households, questions on community life in Putnam County, the desirability of Putnam County as a place to live, social ties with community members and ratings of community life.

Thirdly, I conducted an analysis of expenditure patterns among community members. This survey of households in the county obtained information on how much was spent for goods and services, where money was spent, and why it was not spent locally.

In the following, I explore three broad research questions. First, I examine the extent to which community still exists among local residents. Do residents purchase most of their goods and services locally? To what extent do residents have a psychological identification with Putnam County? To what extent do they have strong social ties with other community members?

Secondly, are there racial differences in the attachment of African-Americans and Whites to the community? Although many of the social institutions have been desegregated as a result of the Civil Rights Movement, the school system and churches continue to be segregated. Changes in the workplace may have brought down some of the barriers between the races.

Finally, the development of Lake Oconee raises several important issues regarding community life in Putnam County. What is the relationship between lake residents and others in the county? To what extent do lake residents participate in local events, organizations and community life?

To examine community life in Harmony in the 1990s, it was necessary to expand the spatial area that was defined in the earlier study. Most of Harmony's institutions and organizations have disappeared (only the African-American church remains) and today there are no services provided locally in the community. Part of the community is submerged today under Lake Oconee (the dam on the Oconee River flooded part of the community). There are very few households left in the territory earlier defined as the Harmony community. I could not find anyone in the area

Fig. 5.1. Harmony, Georgia, in 1941. This is the African-American school that was located in Harmony. At the time, these facilities were better than the school for White students, in part because so few White students lived in or attended schools in Harmony.

who still referred to their community as Harmony. Wilkinson (1991) argues that three essential criteria for a community are: (i) a shared territory; (ii) a local society (individuals can meet their needs through local institutions); and (iii) locality-oriented collective actions. As Wilkinson suggests, however, the shared territory for rural communities has expanded over time as advances in transportation and technology have enabled residents to travel further for basic goods and services and for work. For most of the discussion below we will be referring to Putnam County as the relevant shared territory. There are no incorporated areas outside of Eatonton, which is the county seat.

Community Institutions in the 1940s

At the time of the original study there were separate institutions within the community for African-Americans and Whites. Whites had their own church and elementary school (White children attended high school in Eatonton, a nearby trade centre). The church and school were central institutions for both Whites and African-Americans in the community. Wynne notes that community residents feared the loss of these institutions (Fig. 5.2).

> Residents of the community would consider the loss of the school a calamity to the community. 'When you take the school away you kill the neighborhood.' 'When you take the school out of the community, the community goes

Fig. 5.2. Harmony, Georgia, in 1941. This is the interior of the Jefferson Baptist Church, which was the African-American church in Harmony. This is the only social institution that remains in modern-day Harmony.

to pieces.' A community without a school finds it difficult to attract new residents. Furthermore the loss of the school frequently means of the loss of residents who want their children to have all the school advantages possible.

(Wynne, 1943, p. 41)

There were two Baptist churches in Harmony at the time of Wynne's study, Harmony Baptist Church (White) and Jefferson Baptist Church (African-American). Harmony Baptist Church was established in 1828. Members of the church feared that the decline in attendance would eventually lead to the demise of the church. Many residents blamed the lack of local support for the church on the growing popularity of the automobile. These fears were well-founded as the only remaining local institution in Harmony today is Jefferson Baptist Church. The schools have been consolidated into the Putnam County School System and none of the local voluntary organizations exist today.

By the early 1940s, Harmony already had become dependent on Eatonton, the county seat, for most services and trade. Population decline in Harmony made it impossible for the community to support local merchants. The rise of the dairy industry also forced many Harmony residents to participate more fully in county-wide institutions. For example, almost all of the dairy producers in Harmony were members of the Eatonton Cooperative Creamery, which processed, purchased and marketed dairy products. Though increasingly dependent on services in Eatonton, many Harmony residents had not yet shifted their psychological identification with their locality. The New Deal reinforced the dependencies on Eatonton. There are no other cities or towns in Putnam County, so the primary institutions (school, government, voluntary organizations, etc.) continue to be rooted in the county-wide area. By the 1990s, few residents had even heard their community referred to as Harmony.

Race Relations and Community Life

As Wynne noted, Harmony in the 1940s was two communities, one African and one White. Participation in the community-at-large by African-Americans was restricted to the church and the Masonic Lodge.

> The welfare of the community is the white man's concern they think, not the Negro's, and it is the white man who acts in the interest of it. Even in affairs that might be considered theirs alone, the Negroes behavior is circumscribed by what their white neighbors think about such affairs.

(Wynne, 1943, p. 53)

Although there was a substantial African-American migration out of Harmony after the boll-weevil invasion, African-American institutions did not suffer as much as the White ones. African-Americans remaining in Harmony were strongly committed to the maintenance of local institutions.

This phenomenon is especially interesting given the fact that few of the African-Americans owned any land (only one African-American operator owned land in the community in the 1940s).

Social gatherings of Whites and African-Americans together were considered taboo. Interaction was defined by Whites. In some cases, Whites attended the African-American church, but the reverse was not permitted. The system of social control in the rural South was based on patterned behaviour that controlled the individual in the interest of the dominant White group.

In reading Wynne's account of community life in Harmony, one immediately recognizes the strong interaction between race and class, and the importance this interaction played in influencing the ability of families to respond to the changes occurring in the region. Most owner-operators shifted their dependence on cotton to dairying. Although their level of living had been diminished, owner-operators were relatively successful. Tenants were not able to obtain this security through the transition.

Contemporary Conditions in Harmony

Most Harmony residents participate in county-wide institutions, although this varies, especially among full-time residents and home owners on Lake Oconee. In essence, Putnam County is becoming increasingly divided by lake residence (those who live on the lake and those who do not). Local dependency is what ultimately separates lake residents from others in the community (Cox and Mair, 1988). Many residents are dependent on the locality due to their investments in home ownership. Social ties and relationships also make individuals dependent on their local community for social support and resources. As permanent residents become more socially tied to the locality, they become much more interested in local growth and development issues. Lake residents may never become integrated into the larger community and may fail to develop a shared interest because they are less dependent on the local community for their social and economic life. Seasonal residents may become very attached to their community, through lakes associations and contacts with other seasonal residents. This interaction, however, does not produce a shared interest in the larger community. Thus this type of place attachment does not generate the same interests or attitudes among the two populations. If effect, full-time residents are defining community in production terms and lake residents/seasonal residents define it as collective consumption (Castells, 1983).

Race continues to divide residents, however, especially in terms of church attendance and residence, and in education to a lesser extent. Although the public school system is predominately African-American, a growing number of White children are attending the public rather than the private high school. Whites and African-Americans also are increasingly drawn together in the world of work in the community, especially in the factories that have located in Putnam County over the past decade. African-American participation in the local political system is real but limited. During the course of the follow-up study, there was one African-American on the county board.

As local institutions disappeared in the Harmony community, the primary location for obtaining goods, services and work was Eatonton. Approximately 4–5 miles from Harmony, Eatonton served as the centre of social life in the county. For most Georgia counties, the county seat is the focal point of social, political and economic life in the community. There are 159 counties in Georgia, and most of them only have a few towns or cities in them. Counties are clearly the primary political unit in the state, but they also serve as the social and economic units. School systems are generally county-wide. Thus, for all practical purposes, counties are considered as communities for most residents in rural Georgia.

The population size of Putnam County remained relatively unchanged from the time of Wynne's study in the early 1940s to the late 1970s. Since the 1970s, the population has increased dramatically, growing from about 8000 to approximately 14,000 today. The Putnam County population as a whole grew over 37% from 1980 to 1990 and over 70% between 1970 and 1990. This growth rate is almost twice the state average during the same periods and was nearly triple the US average. The African-American population comprises about 42% of the entire county population, but it is concentrated in the city of Eatonton. City population, however, has begun to decline since the 1980s. This pattern of racial segregation is occurring in other towns throughout the Plantation South (Aiken, 1990).

Fig. 5.3. Harmony, Georgia, in 1941. This is the exterior of a typical tenant house in Harmony, of the type occupied by African-American tenant farmer families.

Putnam County has lagged behind the state and country in terms of educational attainment, and the educational level of African-Americans is significantly lower than that among Whites in the county. In 1990 average completed education in Putnam County lagged behind African-Americans in greater Georgia by over two grade levels, and behind African-Americans in the USA by over three grade levels.

Recent Changes in the Economic Structure

Although there has been change in the economy throughout the past 50 years, the structural changes that occurred in the 1980s were probably the most significant. Manufacturing industry clearly established itself as the basis of the local economy and the economy became much more dependent on absentee-owned firms as the primary source of jobs. Although there continue to be many independent businesses in Putnam County, most serve as suppliers to other large firms in the region.

The early 1980s were a very difficult period for the Putnam County economy. Two major employers, Enterprise Aluminum (450 employees) and Eatonton Manufacturing (350 employees), closed around 1985. By 1987, the unemployment rate in Putnam County was 10.6%, the highest in the state. Just a year later, however, the unemployment rate was down to 5.8% because several new industries located in the county. The largest new employer is Hanes, which employs approximately 400 workers. Approximately two-thirds of the employees hired were from Putnam County. The other new employers during the late 1980s were Gro-Tec, Inc. (fertilizer, 30 employees), BET Enterprises (a subsidiary of Horton Homes, wood products, 23 employees), Cambridge Air Filters (filters, 65 employees), Alonan (aluminium products, 13 employees), Perky Cap (cap manufacturer, 40 employees), Universal Forest Products (roof trusses, 30 employees) and Haband Industries (mail order, approximately 150 employees). The largest employer in Putnam County is Horton Homes, which employs more than 600 workers.

The major factor influencing the location decisions of these new firms was the supply of labour. Because it had the highest unemployment rate in the state for a short period, Putnam County was considered attractive by employers. Interviews with the managers of the new firms in the community confirmed that this was the major consideration for Hanes, Perky Cap, Gro-Tec and Cambridge Filters. The experience of local workers also was important to these businesses. These firms were seeking workers who had experience in sewing factories or in factory settings.

Wage rates were considered the second most important reason for new industry locating in Putnam County. A few managers revealed that they had actually lowered wages since they began operating in Putnam County, not because of the high costs of production, but because the wage levels were lower in the community than they had anticipated. Most of these factory jobs paid between US$6.00 and US$7.00 per hour in the early 1990s.

In most cases, the incentives offered by the community were not a major factor in the location decisions of firms. Availability of land for development was mentioned as a consideration by a few of the firms. This issue was raised by some of the key informants who expressed concern that the city had provided too many incentives for these firms to locate in Eatonton. In particular, the cost of providing the site for Hanes was very expensive. Taxes in Putnam County have risen rapidly in recent years, which has sparked the 'Concerned Citizens Group' to demand tax equalization in the county. The issue of tax breaks to businesses, as well as tax assessments on the lake developments, has been part of the debate regarding tax equalization. Although the community provided some incentives, such as land to most of the new firms, these incentives did not appear to play an important role in firms decisions to locate in the community. In a few instances no other site was considered by the firm. It would appear that the incentives were unnecessary and resulted in a net loss to local government revenues. The 'Concerned Citizens Group' was primarily upset, however, with what they viewed as wasteful spending by the county government and unequal assessments.

There is a growing concern about the direction of economic development in the county. Key informants were all in agreement that the county should pursue industry other than manufacturing, especially 'high tech' firms. Part of the concern is that the supply of low-skilled workers was relatively low. In the words of one person, 'The expansion of the low-wage industries has gone as far as it can.' As another person put it: 'What they did have to sell was a workforce.' There is a recognition that the skills of the existing work force would limit the alternatives for development. New approaches to economic development were being sought by some of the community leaders. One of the sources of the controversy over economic development and land use in the county is the growing division between residents who live on the lake and others in the county. The perceptions that lake residents are not interested in development, while local residents are much more supportive of development efforts.

The Industrial Development Authority Committee is beginning to see the need to be more concerned about the retail sector. Buddy Nolan, who is a lawyer and the Chairman of the Putnam County 2000 committee, pointed out that there are a large number of vacant buildings in the downtown area. Mr Nolan estimated that only about 10% of the residents on the lake are full-time residents, and that few of them have any reason to travel into Eatonton to shop. His general impression is that the city is not benefiting much from the lake development and that this issue should be the focus of future economic development efforts. There may be a greater degree of common interest between lake residents and others over retail development than over industrial development.

Fig. 5.4. Harmony, Georgia, in 1994. This is one of the few tenant houses still remaining in Putnam County, located adjacent to a factory that manufactures aluminium cookware.

To further examine the economic base of the community and how it has changed in recent years, I conducted interviews with most of the major employers in the county. These interviews were semi-structured with owners (and managers) and included questions about the workforce, the community and their business. In the following, I briefly summarize the interviews with each of the major employers.

Major employers

Horton Homes

The largest employer in the county is Horton Homes, a manufacturer of mobile homes. Dudley Horton is in his mid-50s and is considered the wealthiest person in Eatonton. Although his business is considered the 'lifeblood' of the community, he is relatively distant from the community. Starting out in an abandoned chicken house in 1970, Dudley Horton now employs approximately 650 workers, with several small firms operating as full-time contractors. The sprawling complex constructs about 40 homes a day. Approximately 95% of the workforce is male. Nationwide, it is one of the leading companies producing manufactured homes, with an annual sales of about US$100 million. The firm was having some financial difficulty in the late 1980s until it landed a large contract to supply homes for settlements on the West Bank in Israel. Jobs at Horton Homes are considered to be a prize, with a good worker earning as much as US$30,000 per year in the early 1990s. Wages are based on a set of incentives (e.g. attendance, productivity, etc). Horton received state-wide attention a few years ago over his drug testing. He laid off about 25% of his workers after conducting random drug tests. He offered workers failing the test their jobs back if they went into a drug rehabilitation programme.

Hanes

Hanes began its operation in Eatonton in January of 1988. It employs approximately 600 workers when it is at full speed. Robert Edwards, the plant manager, said that new employees go through a 16-week programme to develop sewing skills, but the job is still considered semi-skilled or unskilled. This plant manufacturers underwear for the Hanes line. The decision to locate in Eatonton was part of a strategy on the part of Hanes to locate four plants in Georgia during that year. Most of the plants were located in the North Carolina Piedmont. Hanes, which is owned by Sara Lee, also has 14 plants in the Third World (e.g. Dominican Republic, Costa Rica and Jamaica). Although there is substantial growth in the offshore plants, there are still advantages of continuing production in the USA, such as transportation costs and turn-around time.

Most of the employees hired were from Putnam County (the manager's estimate was 65%). Ninety-five per cent of the employees are female. The starting wage in the early 1990s was approximately US$5.50 an hour. Sixty-five per cent of the employees are African-American and less than half have a high school degree. Hanes has a reasonable benefit package, including health insurance and disability. They recently started an ESOP programme, which is based on 5% of the worker's salary, and the plant recently started a credit union.

Cambridge Filters

Cambridge manufactures filters for commercial and industrial uses. The home office is in Syracuse, New York. Eatonton was chosen because of its high unemployment rate at the time and because it was in a non-union state. The plant opened in June of 1988. The starting wage in the early 1990s was approximately US$4.50 per hour. Among their employees, 75% are from Putnam County, 90% are female, 80% are African-American. At its peak, the company employed approximately 70 full-time workers. Although the firm operated for most of the time during the fieldwork, the home office recently closed this branch down to consolidate operations.

Perky Cap

This firm is owned by Frank Fitzgerald and Jack Perkinson. Perkinson started the firm in the late 1980s. He had been working for a firm in Waycross and decided to open his own business. He was attracted to

Eatonton because of its high unemployment rate in 1987 and by the large number of workers who had sewing experience. In addition, Horton Homes provided the land for the project and the building was built to specification for him. Horton now owns part of the company.

Perky Cap employs about 55 workers. Ninety per cent of the workers are women, 60% are White, and the average education is less than high school. They use a piece-rate system. The average hourly wage is approximately US$5.00 per hour. The starting base salary is approximately US$4.00 per hour, but Fitzgerald mentioned that many workers stay with Perky for less money because they have more freedom than they would at other workplaces. For example, he lets them chew gum and listen to radios. When I visited the factory, employees in the art room were listening to soap operas on the television. Some of these workers had left Hanes where the rules are much more rigid. Perky offers medical and disability insurance, but does not offer life insurance.

What is most striking about the economic opportunities in Putnam County is that there appears to be improving opportunities for White men and both African-American and White women. Fewer jobs are available for African-American men. Most of the sewing jobs are held by women. Women also can find jobs in the growing service sector. The best-paying jobs in the area, for example at Horton Homes, are held almost exclusively by White men.

Dairying

Dairying is the largest agricultural business in the county. There are about 60 dairy farms with an average herd of about 135 cows. Putnam County is the largest dairy-producing county in the state. The gross income from dairy is approximately US$20 million. The largest dairy farm in the county is owned by the Thompson family and includes a herd of about 500 cows.

A typical operation is the Davis operation owned by Heck Davis. The Davis family has a herd of about 250 cows. The operation is a three-family (brothers) operation that was passed on through the family. They do not rely on any outside labour. Other dairy farmers in the community are hiring migrant (Mexican) workers as the labour market is very tight, and they claim it is difficult to find reliable workers. The Mexican community is essentially invisible in Putnam Community, although you sometimes see workers walking along the road. Most of the migrant workers are not citizens and this fact is fairly well known in the county.

The Davis's grow most of their own feed. The Davis farm began in 1948 when Mr Davis moved to Putnam County and began milking 21 cows by hand. Their farm is now incorporated, primarily for inheritance reasons. The Davis farm has good land to farm. The soil actually looks rich compared to that of many other farms in the county and it is not the red colour that is so prevalent in the Piedmont region. They grow most of their own hay and even several acres of row crops.

Forestry

Three-quarters of the land in Putnam County is in forest. The roads in the county are always filled with logging trucks. The gross income from forestry is approximately US$10 million per year. It is fairly common for landowners to sell off some of their land periodically, to generate additional household income.

Community Attachment and Social Bonds

I conducted a survey of Putnam County households to examine their attachment and social bonds in the community. In the survey of local residents, we first asked respondents a series of questions about their community (as defined by them). Respondents were asked: 'If you were hospitalized for two weeks, who besides your immediate family would be willing to do the following things for you?' The list of things included: watch your house, mow your lawn, tend your children, run your errands, lend you money and provide emotional support. We asked Putnam County residents whether neighbours, friends or relatives (outside the home) would perform any of these tasks for them (Fig. 5.5).

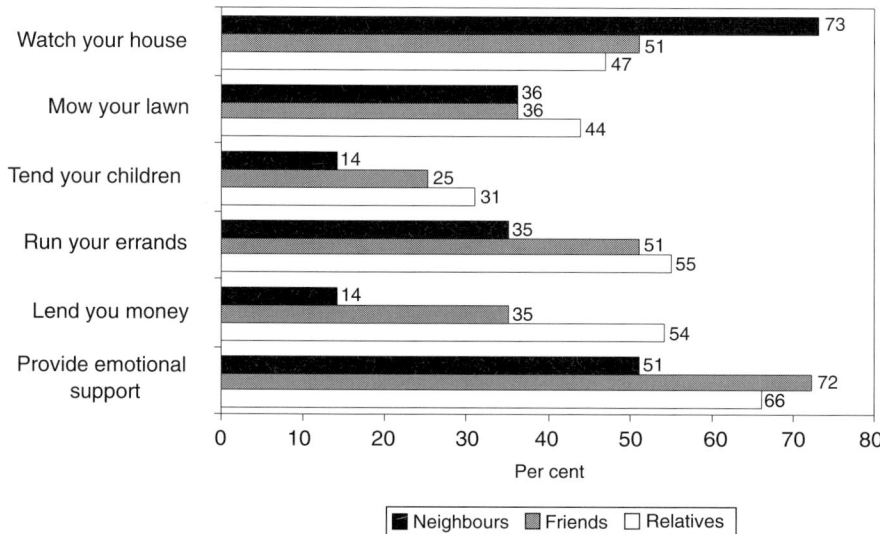

Fig. 5.5. Dependence on neighbours, friends and relatives.

Putnam County residents rely heavily on neighbours, friends and relatives for a variety of things. They are generally willing to ask neighbours to watch their house if they were gone for 2 weeks, but they would not ask neighbours to tend their children or lend them money. Respondents rely most heavily on friends and relatives for emotional support, and on relatives for lending them money.

We did find some systematic differences between African-Americans and White households with regard to dependence on friends and relatives, but not neighbours. In general, African-Americans are not more likely than Whites to rely on neighbours for most of the types of support examined here.

Another indicator of the amount of social interaction among residents is the frequency in which they borrow and trade with neighbours, relatives and friends (Fig. 5.6). Although there are not major differences, Putnam County residents are more likely to borrow and trade with relatives than with either neighbours or friends. African-Americans tend to borrow or trade less with neighbours than do Whites, but more with relatives and friends.

Interviews with key informants and business leaders suggested that the county was considered the 'community' by most residents. Most services are provided county-wide. The school system is county-wide, although many White residents send their children to the local private academy rather than the public schools. Outside of Eatonton, there are few organizations or institutions, although some organizations seem to be emerging in the Lake Oconee area. We asked residents a set of questions about their relationship to the Putnam County area.

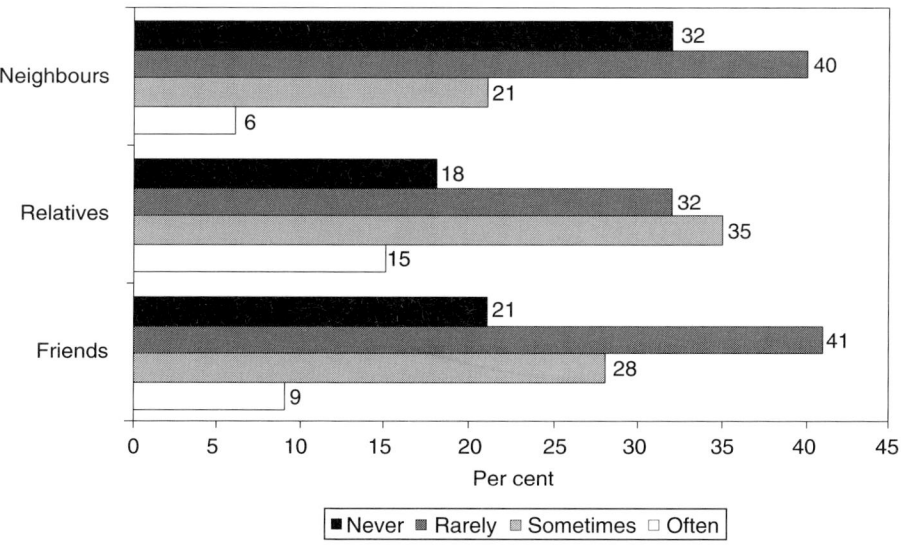

Fig. 5.6. Frequency in borrowing and trading.

Fig. 5.7. Harmony, Georgia, in 1994. This is a home located in the historic area of Eatonton. Efforts to promote the heritage of the community have contributed to restoration and preservation of a number of large older homes such as this.

When asked whether Putnam County is an area in which they would say they belong, the vast majority (83%) indicated 'yes'. And an even higher number reported that they felt at home in this area. Although there is an overwhelming sense of attachment among residents, the level of interest in knowing what goes on in this area varies. Forty per cent of the respondents were somewhat interested, and 58% reported having much interest in what goes on in Putnam County. There are no racial differences in the responses to these questions.

Another indicator of the level of community attachment is the willingness to move from the community. Respondents were asked: 'Suppose that for some reason you had to move away from this area. How sorry or pleased would you be to leave?' Most residents indicated that they would be sorry to leave. Forty-seven per cent said they would be very sorry, and 32% reported that they would be somewhat sorry. Approximately 8% indicated that it would not make any difference one way or another if they moved, 8% would be somewhat pleased, and 4% would be very pleased. Thus, for about 80% of the population, living in the Putnam County area was important for them.

For most Putnam County residents, social ties were rather extensive. We asked residents what proportion of their adult friends, and adult relatives and in-laws lived in Putnam County. Only 4% of the residents did not have any adult friends in the county. For 31%, less than one-half of their friends resided in the county, 22% reported one-half, 37% indicated more than one-half, and 7% said that all of their friends lived in Putnam County.

Residents reported a smaller proportion of relatives and in-laws in the local area than friends. Approximately one-third had no relatives or in-laws in the area. Another one-third had less than one-half of their relatives in Putnam County, one-tenth had about one-half, 16% had most of them in the area, and only 8% had all of their adult relatives and in-laws living in Putnam County.

Overall, there were very strong racial differences in the proportion of friends and relatives that live in the local area. African-Americans were much more likely to have most or all of their friends and relatives living in the area than were Whites. For example, 64% of African-Americans reported that most or all of their friends lived in the local area (Putnam County), while only about 40% of Whites reported the same.

Respondents were asked: 'How many local organizations do you belong to or participate in Putnam County?' Almost all residents reported some involvement. Only 13% did not belong to or participate in any local organization, and 61% belonged to or participated in more than one local organization. There were no racial differences in organizational participation in the county.

To examine the extent to which Putnam County residents rely on local services and organizations, we gave respondents a list of activities and asked them if they usually did these activities locally (Fig. 5.8). In terms of household consumption, residents were most likely to conduct their banking services and purchase groceries locally. Fewer residents purchased hardware (55%), furniture (28%), or appliances (27%) locally. Putnam County residents were most likely to visit friends in the area and attend religious services in the county.

Interestingly there were no systematic racial differences in where residents purchased goods or services or participated in clubs or organizations. The major difference, which is consistent with earlier findings, is that African-Americans were more likely to limit their social contacts to Putnam County than were Whites.

Race Relations

Putnam County continues to be a community divided by race. Although there have been significant advances, African-Americans and Whites live in different parts of the town (Eatonton), attend different churches and have a limited amount of social interaction. Although the public school system has been integrated since the 1960s, a private academy (Gateway School) was established at about the same time. Approximately 15% of the county's students are enrolled at Gatewood. There was a general consensus among key informants that the public school was probably superior to the private academy. Although the public schools are predominately African-American, there are few African-American teachers or administrators.

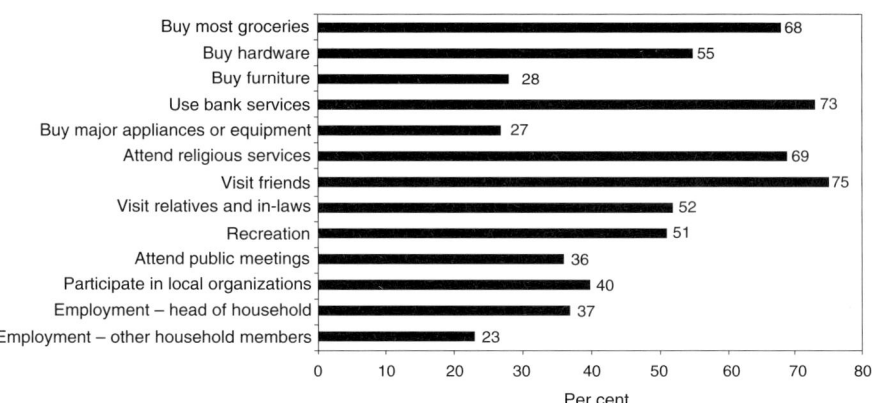

Fig. 5.8. Activities usually purchased within Putnam County.

Although the majority of the population in Eatonton is African-American, the city council is predominately White. One key African-American leader suggested that many African-Americans feel pressured to vote for White candidates. Their employers often encourage them to vote for White candidates. When pressured as to how employers would ever know, she said that many African-American residents believe there are ways of knowing how they voted.

There are three churches in the African-American community: St Johns, Ebineezer and Union. Contrary to many African-American churches in the South, these three churches are not effective social actors in the community. Some African-American residents attribute this lack of effectiveness to the fact that they only have part-time ministers. In two of the churches, services were held every other week. In all three churches, the pastors lived out of town.

There are few African-American-owned businesses in Eatonton. The small African-American middle-class consists largely of morticians and a few small-shop owners, serving primarily the African-American population. The wealthiest African-Americans today are brick-masons, most of whom are in their 50s and 60s. Most of the brick-masons started working for Whites and began their own contracting businesses (Mieher, 1987). Some of them have built their own rental housing in the African-American side of town.

For the African-American community, the lake development has had few benefits. Some of the older women have taken housekeeping jobs on the lake. An informant cited an example of a woman who was making US$5.00 an hour and was paid in cash so she did not pay income taxes. She felt this was an important source of income for many older African-American women in the community who could no longer work in the factories.

The American Legion is probably the most active group in the African-American community. One of the most important events in the African-American community is the Color Purple Ball which is held in early May. The event was named for Alice Walker (the author of *The Color Purple*) who was raised in Eatonton.

The Civil Rights Movement had significant social and political effects on communities throughout the Black Belt. In some counties near Putnam County, the Movement led to radical change. The changes in Putnam County were more subtle, albeit meaningful. As Wilson (1987) suggests, class is probably becoming a more important social cleavage than is race in many communities. The NAACP (National Association for the Advancement of Colored People) still has an active chapter in Eatonton. And every once in a while a racial controversy develops locally, such as the naming of a street after Martin Luther King.

Although African-Americans and Whites share very little in terms of social institutions, they do share a common interest in economic development. African-American and White leaders both talk about the need for better jobs in the community. There is a surprising lack of concern about competition for those jobs between workers of different races.

Lake Residents

Perhaps the most important change affecting social life in Putnam County in the past two decades has been the development of Lake Oconee in

Fig. 5.9. Eatonton, Georgia, in 1941. A busy Saturday afternoon in Eatonton, the centre of trade and government in Putnam County.

1979 by Georgia Power Company. The lake covers approximately 19,000 acres in Greene, Morgan, and Putnam Counties. In addition to providing housing for part- and full-time residents, approximately 150,000 visitors use the public recreation areas each year. The lake development has not produced a significant increase in year-round residents, and one recent report suggests that Lake Oconee may have had a depressing impact on population growth over the past decade, because of persons who were displaced when land was purchased to form the lake bed. More than one-half of Lake Oconee residences are in Putnam County, which had 41 subdivisions with about 800 residences in the early 1990s. Greene County has about 570 residences in 43 subdivisions and Morgan County about 180 in 18 subdivisions.

There are three major developments on the lake: Port Armor (which was purchased by a British firm), Reynolds Plantation and Harbor Club. All three focus on higher-income residences. Several golf courses were constructed – one by Jack Nicklaus at Reynolds, the other by Tom Weiskopf at Harbor Club. One golf course at Reynolds Plantation is ranked among the ten best resort and public courses.

Land prices in the region have skyrocketed in recent years. Ten years ago lakefront lots were going for US$13,000, and in the early 1990s they sold for almost US$200,000. Home prices range from under US$200,000 to a half a million dollars or more.

Among the residents of Lake Oconee, approximately three-quarters are part-time and one-quarter are full-time residents. More than half these residents purchased their property in the past decade. The educational level of lake residents is relatively high, with a quarter of the men and one-fifth of the women attending either graduate or professional school. Thirty per cent of the lake residents have total family incomes exceeding US$75,000 per year.

A survey of purchase patterns for goods and services among lake residents reveals that part-time residents frequently purchase groceries, drugs, beauty supplies, garden and building supplies, boats and marine supplies, hardware and liquor in the Lake Oconee area. Part-time residents rarely shop locally for clothing. Most part-time residents bring what they need for their lake home from their primary residence. They will purchase clothing, furniture and appliances in Atlanta rather than the lake area. Full-time residents continue to purchase many of these same goods in Atlanta rather than locally. It should be pointed out, however, that many residents not on the lake also purchase goods and services outside Putnam County, primarily for price, selection and perceived quality.

A growing division within the community is between lake residents and other residents in Putnam County. The split is often manifested in debates over taxes and zoning. New residents are generally much more in favour of zoning than are the natives. In addition, new residents are requesting more and better services from the county. However, the zoning issue is the primary one facing the county now. One informant suggested that the county was reluctant to pass any zoning ordinances, particularly regarding trailer (mobile) homes because Horton Homes, the manufacturer of trailer homes, is the largest employer in the county. As a member of the Industrial Development Association reported, a number of neighbouring counties had passed zoning legislation and as a result 'we are getting all of their [Morgan and Greene Counties] trash'.

In some cases, lake residents have developed their own organizations, such as the 'Newcomers Club' which was started by women living on the lake. Because of the social divisions between lake residents and others, most lake residents do not feel wanted or comfortable in the Putnam County organizations and institutions.

Conclusions

Putnam County's economic base has undergone a significant transformation since the 1940s. Although dairy production continues to be an important source of income, the manufacturing sector has become the primary source of employment in the county. Tourism and recreational homes also have become important to the local economy. These changes have had several important implications for the region. First, the shift in the economy has integrated more fully the local economy into global markets. One of the consequences has been the emergence of major swings in employment conditions in the region. Secondly, manufacturing employment has provided more opportunities for African-Americans. Although wages are still relatively low, it has provided a more stable source of employment for low-wage, low-skilled workers. Finally, the growth of manufacturing employment has increased the interaction among African-Americans and Whites (more for women than for men, however). Although African-Americans and Whites continue to live in segregated neighbourhoods, they are much more likely to work together. Thus, among most residents who do not live on the lake, there is a consensus that economic development is beneficial to the community.

Over the past 50 years, Putnam County has experienced major changes, including the rapid growth of non-farm employment, the Civil Rights Movement and rapid economic development along its lakes. Although economic opportunities have improved for many African-Americans, the community remains divided along racial lines. Separate institutions persist. African-Americans are not necessarily more attached to their community, but their family and friends are much more likely to be in the local area.

Other divisions have developed in the community, which have produced new tensions. The most prominent conflict is between lake residents and long-term natives of Putnam County. Lake residents (either full-time or

Fig. 5.10. Eatonton, Georgia, in 1994. Downtown Eatonton on a Saturday afternoon. In stark contrast to the image from 1941, there are few people on the streets, reflecting the reduced levels of downtown business activity in contemporary times.

seasonal) have little interaction with other Putnam County residents. Lake residents are establishing separate organizations and institutions. This finding supports some of the community sociology literature that suggests that these residents would develop separate organizations and institutions to meet their needs (Marsden et al., 1993).

Harmony is much less of a community than it was 50 years ago. Because of its proximity to several metropolitan areas and improved transportation systems, many residents work, and purchase goods and services, outside Putnam County. Fifty years ago, most residents worked, lived and consumed in the same territory. Relatively few Putnam County residents regularly attend public meetings or participate in local organizations today. But most residents continue to maintain social ties in the local area and purchase most of their goods and services locally; and most residents would be very sorry if they had to move from the area.

In addition, there continue to be issues that spark locality-oriented collective action. Two recent issues illustrate this point. The 'Concerned Citizens Group' was a fairly broad-based effort to promote tax equalization and a more responsive county government. The other example is the effort to increase the availability of good jobs in Putnam County. The Chamber of Commerce and the local development organization have generated community-wide support for their activities in recent years.

Like Waller Wynne, I am also taken by the persistence of community life in Putnam County, despite major changes in the economy and culture of the region over the past 50 years. Although race continues to be an important division in the community, the movement of manufacturing firms to the region has provided increased opportunities for African-Americans. The New Deal programmes and the Civil Rights Movements also improved opportunities, although African-Americans continue to lag far behind Whites in almost all measures of economic well-being. New divisions, however, are emerging as the economy is once again shifting. Growth of tourism, recreational homes and a post-industrial economy present new obstacles to community in Harmony.

Note

[1] This material is based upon work supported by the Cooperative State Research Service, US Department of Agriculture, under Agreement No. 91-34229-6077. Any opinions, findings, conclusions, or recommendations expressed in this publication are those of the author and do not necessarily reflect the view of the US Department of Agriculture. The Kellogg Foundation provided additional support for this project. I appreciate the assistance of Deborah Klinko Cowell and Tsz Man Kwong in the data collection for this project. The residents of Putnam County were more than generous with their time and hospitality. Ken Wilkinson provided insightful advice and guidance early in the project. Finally, the other researchers on NE-173, Community Change and Persistence: A Restudy of the Rural Life Study Series, have played an important role in this research. Len Bloomquist, Eric Hoiberg, Fred Schmidt, Al Luloff, Rick Krannich, Clyde Eastman and Lou Ploch have been valuable colleagues in every sense of the term.

Community Change and Persistence: Landaff, New Hampshire

Fred Schmidt, Elizabeth Skinner, Louis A. Ploch and Richard S. Krannich

Introduction

This investigation focuses upon how the small town of Landaff, New Hampshire, has adapted to and withstood change over the last 50 years of the 20th century. It is a study of community persistence in the light of sometimes dramatic social, economic and cultural changes. This study examines Landaff's environment, agricultural and economic development, local perceptions of the community, residents' sense of place and belonging to the community, as well as the ways in which they define their community, and other social characteristics, including levels of social participation and interaction.

As noted elsewhere in this volume, the communities selected for the original USDA-sponsored Rural Life Study series were chosen to represent a continuum from high to low stability. When the original studies were initiated in 1940, Landaff was deemed to be a 'middle' community on that continuum. However, events of the intervening half-century suggest that those who developed the stability–instability criteria may have underestimated the importance and persistence of local government, especially in northern New England communities. In this region, despite many external changes, local self-determination focused primarily on local affairs continues unabated. The persistence of local government and, correspondingly, local social organizations lends communities in this region a stability atypical of many of the other study communities found in other areas of the USA.

When Kenneth MacLeish and Kimball Young (1942) reported on the social and economic conditions of Landaff, they found an agricultural community that was perhaps not as stable as previously thought, but also not as severely buffeted by the forces of change as might have been expected. At that time, Landaff was a community characterized by small family farms, all caught in the struggle of adjusting to a transition to commercial dairy activities within the larger regional market. Few local farms were able to compete in this newly emerging environment. As a result of the declining economic opportunities, many of Landaff's younger generation were forced to leave town in search of employment. This out-migration weakened, but did not destroy, the community's culture or its adherence to values defined by the old-time farming families, which was manifested in an emphasis upon independence and hard work. MacLeish and Young attributed Landaff's continued existence to both Yankee ingenuity and the ability to adapt to changing situations without sacrificing core values.

In the late 1980s, Louis A. Ploch (1989), Professor Emeritus of Rural Sociology at the University of Maine, revisited Landaff. He found a community that, despite its continued lack of economic vitality and correspondingly long-term patterns of decline associated with dairy farming in the area, was still able to be classified as being midway on the stability–instability continuum. Ploch (1989) noted, in particular, that substantial changes in land-use patterns had occurred as lands previously dedicated to farming were redefined as highly desirable areas for rural residential development. This change exhibited the eternal nature of Landaff: it would not die because it was attracting a new breed of in-migrants–retirees and younger professional families in search of a rural way of life. Indeed, such in-migration served to counterbalance the continued out-migration of Landaff's youth, seeking external employment.

Community, the critical unit of analysis in this chapter and the key organizing concept for this book, is a time-honoured sociological construct and one of the discipline's fundamental conceptual units. Definitions of community abound, to the point that it sometimes seems there are only slightly fewer definitions than there are individual communities. Seminal work by George Hillery in the 1950s reduced the plethora of definitions to critical 'themes'. Legitimized by his work (and that of others), contemporary community sociologists typically focus attention on six distinct themes. These six are: *size* (more than two and 'under' a debatable upper limit of inhabitants on an extant, but not necessarily fixed, land or water base); *longevity* (sufficient persistence to exceed the normal life of any single member); location in *relationship to a natural resource base*; exploitation and/or *manipulation of that resource base* (takes shape as opportunities for both the local economy and its recreational activities); *ties of sentimentality and solidarity* (albeit a collective emotion with an ebb and flow over time); and *continuity in decision making* (local polity, economy, religion and education).

These six themes are evident in much, but not all, of the community literature. When MacLeish and Young focused their study (1942) around six broad categories, they identified similar themes: identification and characteristics of the community; history and background of the community; making a living; the community, organization and values; impact of the outside world upon Landaff; and integration and disintegration in community and individual life.

The analysis presented here synthesizes these major thematic categories into three core themes: community characteristics; making a living on an identifiable resource base; and community identity, organization and solidarity. In addressing the last of these themes we focus particular attention on the role of local government, a factor of central importance in understanding the patterns of change and persistence in this New Hampshire community.

Study Approach

Previous researchers have used a variety of techniques to inform their portrait of Landaff. MacLeish and Young, for example, utilized participant observation as their primary method. MacLeish spent 4 months living in Landaff, followed by a series of subsequent visits. Young visited the community to conduct interviews and was responsible for comparing Landaff to the other Rural Life Study communities with respect to societal and cultural trends (Taylor, in the foreword to MacLeish and Young, 1942). Their observations were supplemented with data from sources outside the community, including census records.

Ploch's (1989) restudy of Landaff was based on data collected through interviews of key informants. By initially interviewing personnel selected from various county agencies, including the extension service, he obtained information about the community and identified appropriate local contact people. He then held interviews with 36 full-time Landaff residents, one vacation-home owner, five state/federal agency personnel whose territory encompassed Landaff, and three Lisbon, New Hampshire school officials whose district educates the majority of Landaff's older children. Among those interviewed were present and former selectmen, school board members, fire chiefs, chairpersons of the planning board, and present and former farmers.

The early-1990s study focused on here was conducted by Dr Fred Schmidt and his associates in the Center for Rural Studies at the University of Vermont. The key techniques used in that restudy were secondary data analysis, field research and a resident household survey. Analysis procedures included: (i) an in-depth review of the two previous community studies (MacLeish and Young, 1942; Ploch, 1989) and of a detailed historical report on Landaff's first 200 years (Currier and Clement, 1966); (ii) a secondary analysis of demographic and land-use statistics; (iii) a self-administered mail survey of all households in Landaff; and (iv) field visits and interviews with key Grafton County and Landaff government personnel. In addition, aerial photographs from 1955, 1982 and 1994, and geological survey maps from 1967 were reviewed. Although this information was critical for indicating broad patterns of land-use change, because those data were incomplete they are not formally reported here.

This chapter relies most heavily upon data obtained from the household survey. A self-administered mail survey instrument (developed as part of the restudy effort by all of the contributors to this book) was delivered to all Landaff households in 1993. The major foci of the survey were the measurement of community attachment and participation. A mailing list of local residences was compiled through local property ownership records and amended through examination of a list of local voters and, subsequently, current phone listings. The initial list, which was reviewed by several knowledgeable local leaders, established the presence of 152 residences (despite the fact that the most recent census had reported 139 households). Field reconnaissance and consultation with local leaders indicated that several homes were either uninhabited or only intermittently inhabited by seasonal residents, reducing the final mailing list to 142 households. Emulating the design employed in other study areas, which sought one response from each household, the survey targeted the (self-defined) household head, adult spouse or other adult over the age of 18 with household decision-making responsibilities. Three separate first-class mailings were conducted in order to increase the survey response rate. Despite these efforts, the yields from the second and third mailings were disappointing. Over all, 12 survey packets were returned due to inaccurate or incomplete address information, and six were returned indicating that the addressee was deceased or unknown. This resulted in an effective final list of 124 viable addresses to which survey packets were delivered. From these a total of 78 surveys were returned, yielding a response rate of approximately 63%.

Community Background and Population Characteristics

Landaff has been populated continuously by Anglo-American residents since at least the 1760s; inhabitants of the local area first petitioned for formal community status in 1764. When MacLeish and Young conducted their study in 1940, they described Landaff as a small hill community (elevation 1144 feet) in Grafton County, located approximately 150 miles north-west of Boston in what is known as the North Country region of north-western New Hampshire.

As Ploch (1989) observed, Landaff's remote location, limited soil fertility, paucity of level land and other requirements for industrial development, and lack of attractive fishing, lakes or mountains suitable for commercial winter recreation were detriments to both in-migration and population retention. Despite the fact that Landaff's history has been linked to its agricultural roots, MacLeish and Young reported that 38% of the soils in Grafton County were unsuitable for agriculture. The remaining soils were considered to be sufficient for subsistence farming, but not for commercial production. New Hampshire's climate is also quite harsh, with the exception of the summer months: the average annual temperature is 42°F (5°C), the January average is 15.6°F (−9°C) and the July average is 66°F (19°C). While a reliable annual rainfall (annual average precipitation of 38 inches) was an asset to the community, late springs and early winters meant a short growing season. Such conditions proved increasingly problematic with the onset of commercialization and competitive agriculture. Given these circumstances, Ploch (1989, pp. 13–14) found it somewhat surprising that Landaff's population didn't decrease faster than it did during the years following the Great Depression.

However, when evaluating the patterns of change that have affected Landaff, it is important to remember that the typical community in this region of the country (especially in the northern New England states of Vermont, New Hampshire and Maine) has legally mandated local responsibilities. Local jurisdictions enjoy powers of taxation and decision making that were granted during their chartering process over 200 years ago; these powers have been honoured through the continuous service of citizen-residents. Such 'civic responsibilities' are passed from one generation to the next. When that pattern is threatened through dramatic in- or out-migration, the legacy of voluntary community service often entraps the current generation of established residents in an expectation that they will stay in place and fill these offices, and continue to make important decisions; as a result, local control will continue to be exercised. Such traditions die

Fig. 6.1. Landaff, New Hampshire, 1942. The Blue School.

very slowly in these states. Indeed, very few towns have dis-incorporated in New Hampshire or elsewhere in northern New England over the 200 years of their history. The overwhelming majority of incorporated places have persisted over the years, even when faced with severe economic and demographic decline. Unlike states to the south and west, persistence of community is a given in northern New England.

Landaff was included in the original Rural Studies series investigations because it was an old community that had experienced a long period of stability, but where one critical sector of the local economy, dairy production, had been considerably disrupted in the immediately preceding years by the development of new economies of scale associated with the larger Boston milkshed (Taylor, 1942). MacLeish and Young (1942) reported that this threat to the continuity of small dairy farms was a major influence on the social and economic conditions evident in the community. Taylor (1942) concluded that MacLeish and Young's study cast doubt on the original assumption that Landaff had a high degree of stability. However, these interpretations underestimated the stability of the local form of government in New England. Local decision making through school and select boards and, increasingly, planning and zoning commissions, recreation and conservation and/or natural resource boards has remained remarkably consistent over time. Each of these functions typically enjoys jurisdictions co-terminous with community boundaries, and thus represents a pattern of social organization that is unique in its stability across a span of decades and centuries. At least in retrospect, New England towns are remarkably stable; barring disastrous population decline and sustained out-migration, it is very unlikely that they will disappear as civil units.

Landaff's population has never bounced back from its post-Civil War decline, although some growth has been experienced since 1980. At the time of the 1940 study, the community's population was 389 – population loss continued at a steady pace for the next 40 years. At the time of MacLeish and Young's study, the biggest challenges facing Landaff were the decline of agriculture and changing land-use patterns. While Landaff had always been a farming community, by the 1940s farming had declined significantly as the push to commercial competitiveness led to a focus on commercial-scale dairying. This trend continued in the 50 years following the 1940 study, with fewer farms and fewer people involved in an agricultural way of life. Whereas in 1940 half of the population (195 of 389 people) resided on farms, data from the 1990 decennial census indicated that fewer than 10% of residents (34 of 350 people) were classified as a rural farm population.

As was common in much of rural New England during the 1970s, Grafton County exhibited a nearly 20% population increase from 1970 to 1980. Meanwhile, Landaff continued to lose residents, declining by nearly 9% in population over the same period. However, Landaff's population decline began to turn around during the 1980s, when the number of residents increased by nearly 32% – more than twice the rate of change experienced in Grafton County during the same decade. During the first half of the 1990s, population growth tapered off in both the county and the town, with Landaff exhibiting just below a 1% increase in population between 1990 and 1995.

Ploch (1989) identified several reasons for Landaff's inability to exhibit the population growth that was occurring in other northern New England locales prior to the 1980s: (i) relatively low birth rates and high death rates in an ageing population; (ii) continued out-migration of young people; and (iii) Landaff's location in the far western portion of New Hampshire, which meant that it was well out of the reasonable commuting range of the greater Boston metropolitan area.

Despite the effects of out-migration and lack of economic growth, Landaff exhibited a relatively high level of stability with respect to the population that did remain in place. According to 1990 census data, more than 78% of Landaff residents resided in the same home as in 1985. Of the remaining 22%, nearly 9% relocated to Landaff from other communities in Grafton County, 5% from other places in New Hampshire and nearly 8% came from out-of-state. According to findings from the 1993 Center for Rural Studies survey of Landaff residents, the mean length of residency was 20 years (with a standard deviation of 19.6 years), with a maximum of 83 years. Along with the above-mentioned influence of local government institutions, this continuity of residency among a large portion of the local population contributes to the persistence of community organization.

Census figures from 1990 indicated that per-capita income among Landaff residents was US$11,942 and the median household income was US$31,607. The median age was 37.3 years. Of Landaff's citizens aged 25 years and older, 74% were high school graduates and 18.5% had received a college degree. While other areas of the country have seen an influx of foreign immigrants and 'ethnic' populations, Landaff continues to be primarily White and of Yankee stock. In the 1990 Census, Landaff's racial make-up was White except for one woman between the ages of 65 and 69 who was categorized as Black. The 1990 census also documented that 100% of the population was native-born and 59% were born in New Hampshire, with the majority coming from English, French (French–Canadian), Irish or German ancestry. This is in contrast to the 1940 census, in which 6.17% of the population were foreign-born (24 of 389 people), although all were White.

Another useful indicator of community change is the proportion of housing units used on a seasonal or recreational basis. According to the 1990 census, there were 139 occupied housing units and 57 vacant units. Of the vacant units, nearly 75% (42 units) were used on a seasonal, recreational or occasional basis. In other words, more than 20% of all of Landaff's housing units were used on a seasonal or recreational basis, compared to just a small handful of houses that were rented out to summer residents in 1940 (see MacLeish and Young, 1942, p. 48).

When Ploch revisited Landaff nearly a half-century after the United States Department of Agriculture (USDA) study, he found that while Landaff remained a relatively stable community, there were indicators of significant changes both in residence patterns and, correspondingly, in land use (Ploch, 1989). Landaff's rural setting has now become an asset and has attracted a new breed of in-migrant. The latter tend to be retirees and younger professionals with families who either commute to non-agricultural jobs or who are self-employed and work from home. Landaff today is a mix of elderly long-time residents and relative newcomers of varying ages. While the changing composition and increased size of the local population certainly are altering the social dynamics of community life, they also exert a stabilizing influence on the community by helping to maintain the population base needed to support key local institutions.

Making a Living: Community Resource Base

When Anglo-American settlers first occupied north-western New Hampshire they found a natural resource base – soil, climate and topography – that was well suited to small, family-oriented animal farms. Not surprisingly, it was this type of farm that dominated the landscape throughout the 1800s and into the 1900s. From Landaff's early history to the pre-Second World War era, nearly all families were involved in agriculture, forest production or a combination of the two (Ploch, 1989, p. 12). Initially, local agriculture was quite diversified and many crops were produced for

Fig. 6.2. Landaff, New Hampshire, 1995. The Blue School today looks much the same as it did 50 years ago.

the family's needs or for barter with local stores. Landaffers produced maple sugar, butter and cheese, and raised sheep (primarily for wool). Potatoes were grown in the latter half of the 19th century for starch, and, at one time, there were as many as three starch factories along Mill Brook in Landaff (Currier and Clement, 1966). Much of this starch was dried in kilns and sold to print works in Massachusetts, or used in the manufacture of cotton, wool and linen goods.

The introduction of the railroad in 1854 allowed Landaff to join in the process of commercialization (MacLeish and Young, 1942, p. 100). More products were brought to market, and factories were established in the area. Lumbering had already been profitable in Landaff's early days, both for wood and spruce oil, a compound used in insecticides, perfumes, shoe shine and medicines. Wool became a profitable business during the Civil War period, since cotton was in short supply in the USA.

As agriculture throughout the nation became increasingly specialized and technologically driven, the climate and altitude of Landaff dictated an agriculture shift to the only area in which it could maintain a competitive advantage – dairying (MacLeish and Young, 1942, p. 6). As Carl Taylor noted in his foreword to MacLeish and Young's study, the industrialization of Landaff's agriculture was triggered by the penetration of the Boston milk-shed into Landaff. This industrial approach to dairying was marked by an increased emphasis on milk production levels, sanitation, housing, breeding and feeding, as well as a shift from a barter system to monthly milk cheques, and from general self-sufficient farming to commercialized dairying. This shift also produced cultural changes, transforming Landaff from a *gemeinschaft* community (informal, personalized) to a *gesellschaft* society (rational, impersonal). The need to produce enough milk to pay for the new inputs required by industrialized dairying became the primary goal, and 'those persons who could not, for whatever reasons, make the transition … had to leave town or find other ways to make a living' (Ploch, 1989, pp. 12–13).

Because local jobs were scarce, out-migration and consequent population loss became part of Landaff's heritage. The number of farms decreased significantly, as land from small farms that no longer survived was rented or bought by larger farms, many of which tended to be specialized dairy farms. Even by the 1940s, more than one-half of Landaff's acreage (mostly forested) was owned by non-residents and non-local organizations, including a lumber company, the federal government, a bank, and two or three individuals (MacLeish and Young, 1942, p. 29). Of the remaining land, 98% was held by 36 of the 72 resident families in Landaff. The holdings of the farmers varied from 60 to 1000 acres, with most between 100 and 200 acres. Long-time Landaff families owned most of the larger holdings (Fig. 6.3).

At the time of MacLeish and Young's study there were 35 operating dairy farms in Landaff; by 1988 only two dairy farmers were still operating

Fig. 6.3. Landaff, New Hampshire, 1942. Landaff Town Hall – home to major municipal committees and town meetings.

(Ploch, 1989, p. 20). While some small-scale timber harvesting continues in the area, the forest products industry is no longer a major source of employment or income for the local population. At the end of the 1980s, no manufactured goods were produced locally, and there were no retail outlets, restaurants, commercial sales operations or other similar forms of commercial activity and income generation (see Ploch, 1989, p. 26). Indeed, by the time of the Center for Rural Studies restudy in 1993, it could be argued that Landaff, *per se*, exhibited little in the way of a local economy, at least in the traditional sense of local economic production activity.

Like many other rural communities across the USA, Landaff has evolved from a place with its own distinct local economy to one where economic activity and the livelihoods of local residents are enmeshed in the circumstances of a broad area and in regional, national and even global contexts. In recent years, the arrival of substantial numbers of new residents, many of whom are employed in professional, managerial, sales and services occupations in surrounding and even distant communities, indicates the economic values associated with natural amenities and 'rural ambience' (see Ploch, 1989, p. 26; also McGranahan, 1999). In many ways, Landaff has become a bedroom community for persons whose preferences to live in a rural, small-town setting counterbalance the absence of local economic activities or opportunities, and who either can rely on income sources not requiring local-area employment or are willing to commute a considerable distance to jobs outside of the local community. Ultimately such patterns can give rise to a fundamentally different local economy than has traditionally been evident in rural America. Amenity-based in-migration often is associated with increased land values, added local tax revenues, increased demands for public as well as private sector service provision, and increased public and private sector employment opportunities. While Landaff has little prospect of returning to a commodity-based local economy, it appears to be moving toward the development of these kinds of amenity-based economic activities.

The 1990 census reported that of the population 16 years and older, 64.8% were in the labour force. Of these people, 95% were employed (181 people). Broken down by gender, 75% of the males and 54% of the women were in the labour force and either working or eligible to work. Overall, more than one-third (35%) of adults (103 people) 16 years of age or older were not in the labour force, a figure that includes retirees. Of those responding to the 1993 survey, 54% were employed, 1% unemployed, 35% retired and 10% 'other'.

During 1989–1990, Landaff's commercial services consisted of two milk-shipping farms, an auto repair garage, a logging company, a septic tank service, a construction and excavation company, an upholstery repair shop and a junkyard. Most working Landaffers (81%) were employed outside the community. Census data from 1990 indicated that nearly one-third of Landaff residents (31%) were employed in the manufacture of durable and non-durable goods, followed by employment in construction (16%). Agriculture, forestry, fisheries and mining employed fewer than one in ten people (9%). Sixty-two per cent of workers were classified as private wage and salary workers; nearly 19% were local, state or federal government workers and the same percentage were self-employed.

In 1940, MacLeish and Young recorded occupations by family. While we cannot make equivalent comparisons with the 1990 data, it is interesting to look at the percentages to see how industry/employment structure changed in Landaff over the ensuing 50 years. In 1940, 56% of Landaff's families were involved in agriculture or forestry businesses. In 1990, fewer than 10% were in agriculture, forestry or fishing, while the majority of Landaff residents worked in machine operating, assembly and production occupations.

Community Identity, Organization and Solidarity

Landaff's primary formal social organizations in the 1940s were the Grange, the 4-H Club, the Epworth League, the Home Demonstration Club and the Methodist church and its affiliated groups such as the Ladies' Aid Society. MacLeish and Young noted that only about half of the town's families participated in any of these organizations, a fact that may have reflected the diminishing role of agriculture and/or a lack of active community involvement by the retired and the young day-labourers. However, despite the decline in farming, the Mount Hope Grange, organized in

1861, was one of Landaff's most enduring and important institutions and remained the centre of social activity in the community at the time of the 1940 study. The Grange sponsored a number of community events, including Landaff's Old Home Day, the most important community event of that time, and one that continues to the present. Despite its focus on agricultural issues, the Grange was more a social than an educational or political organization in Landaff, and was certainly instrumental in creating a feeling of solidarity among local residents.

MacLeish and Young observed two major forms of social interaction in Landaff in the early 1940s – informal association and cooperation. Cooperation was not social, *per se*, but was an important form of interaction among farming families who provided mutual assistance with farm work, machinery use and repair, etc. At the time of the original study there was not much in the way of recreation and informal association, as finding time to socialize was difficult for most involved with farming. Visiting was an infrequent practice. Families were isolated during the winter months when the roads were bad, but did visit more often during the summer months, especially in the evenings and on Sundays. Dances were held once or twice a year in Landaff, but were offered more frequently in neighbouring towns.

MacLeish and Young found that Landaff's 72 families (excluding some in outlying areas) thought of themselves as a group distinct from other groups:

> The sense of place or distinctive locality is of great importance; it characterizes all of the farm people and even the nonfarm group who have lived in Landaff for many years. It is maintained and strengthened by the township governmental system, whereby members of the community assert their membership in it and their right to take part in its affairs. But aside from this political phase, Landaff is a symbol or value as a point of identification.
>
> (MacLeish and Young, 1942, p. 106)

MacLeish and Young noted that an important non-institutional factor that reinforces community solidarity is the sense of legacy – the feeling of belonging to a place, to a piece of land, and of having descended from people who lived on that land long ago. They noted that Landaffers expressed a 'feeling of belonging and a consciousness of difference from those who do not belong' (MacLeish and Young, 1942, p. 4). Even in the face of substantial and sustained out-migration, a sense of group solidarity and identity was clearly evident.

Local schools played an important role in promoting community identity and solidarity. While population declines in prior years led to the closure of some of the seven schools that at one time had served the community, in 1940 there were still three local school districts – Scotland, Ireland and the Blue School – each with its own one-room schoolhouse. MacLeish and Young reported that the schools played an important role in the community. The children almost exclusively associated with other community children, and school management and operations issues brought the adults together. They also found that the community's farm families controlled local education as they did local political activities, even though by 1940 non-farm children outnumbered farm children.

MacLeish and Young also noted that Landaff's local government provided the clearest expression of solidarity and self-determination in the locality, noting that 'no other influence [had] so definite an effect in maintaining the community' (MacLeish and Young, 1942, p. 78). Town meetings were both political and social events that were attended by entire families, and it was common for the town women to prepare dinner for the assembled group. Town government – like those of other New England communities – operated in nearly the same manner as it had for 200 years, with management of local affairs carried out by three selectmen elected at the annual town meeting on dates specified by state statute. Elected positions were most often held by farm people, reflecting the class structure of the community. At the time of the 1940 study, Landaff's government was in the hands of men who fulfilled the traditional qualifications of high status and local parentage (MacLeish and Young, 1942, p. 80).

Much of the independence and attachment to the land that MacLeish and Young observed is still present in Landaff. Many residents have chosen to remain in, or relocate to, Landaff despite the lack of economic opportunities and the decline of farming. When survey respondents were asked why they continued to live in the area, nearly half (47%) said that it was because of the rural area and lifestyle, and nearly one-quarter (22%) said it

Fig. 6.4. Landaff, New Hampshire, 1995. The Town Hall, like the Blue School, remains a central feature of community life in Landaff.

was because they had a legacy there. When asked how sorry or pleased they would be to move away from Landaff if forced to, 66% said that they would be very sorry to leave. Factors mentioned by respondents in explaining this feeling included its rural setting (41%), its legacy and sense of home (33%) and its people (26%). Bivariate analysis revealed that while age and educational attainment were not significant factors in accounting for whether or not people would be sorry to leave Landaff, those who had lived there for 10 years or longer were significantly more likely to indicate that they would be sorry to leave than were shorter-term residents.

Length of residence in the community is also a significant factor in determining why people continue to live in Landaff. Those who responded that work (4%) was the reason they continued to live there had the highest mean length of residence (39 years) in the community. Those who cited having a home or legacy in the community (22%) had the next highest mean length of residence (31 years). Those who cited the rural setting (47%) had lived in Landaff for an average of 15 years. Those with the shortest mean length of residence claimed that they remained in Landaff for a variety of other reasons, such as the soft housing market.

The majority of survey respondents, especially retirees (63%) and those who were employed by companies or businesses (48%), listed 'being fond of rural areas' as the main reason they lived in Landaff. The majority of self-employed people (30%) listed work as their reason for living there.

About one-half of respondents to the 1993 community survey (49%) indicated that the area they considered as comprising their community encompassed an area larger than Landaff. This may be attributed to the fact that Landaffers receive public education beyond elementary school, as well as most other services, from neighbouring towns, such as Lisbon and Littleton. Many listed Landaff and Lisbon as being part of their local community while others included nearby towns such as Lyman and Littleton (Conner and Schmidt, 1996). A majority (39%) indicated that the services they received and/or their social activities defined their sense of community, although many Landaffers also defined their community as the place where they had a legacy (17%) or a sense of home (16%). Survey participants cited Lisbon as the place where most residents did their banking and went for auto repair services (see Table 6.1). Littleton was cited as the place where residents most frequently obtained groceries, medical and dental care, hardware and home furnishings and major appliances.

Gender influenced how respondents defined their community. Sixty-seven per cent of men responding to the survey indicated that the area comprising their neighbourhood extended to an area larger than Landaff, compared to 40% of the women. The majority of men (68%) cited 'receiving services' as a primary basis for defining the geographic configuration of their community. In contrast, women's responses were split across several categories, including receiving services (23%), a sense of community legacy (21%), fondness for the rural area (14%), the presence of friends

Table 6.1. Local towns and available services.

Services	Littleton	Woodsville	Lisbon	Other	None
Groceries	√				
Medical care	√				
Dental care	√				
Hardware	√				
Banking			√		
Home furnishings	√				
Auto repair			√		
Religious services					√
Recreation & entertainment				√	
Farm inputs					√
Clothing				√	
Housewares		√			

(14%) and familiarity with the area (7%). Interestingly, none of the men responding to the survey listed fondness, friends or familiarity as criteria used in defining the area that they viewed as their neighbourhood/community.

Seventy-four per cent of the respondents said that they feel very much at home in Landaff. Again, connections to other people and a sense of legacy were the most common responses for why Landaffers felt at home (31% each). Twenty-one per cent claimed that the rural setting was responsible for how at home they felt in Landaff. Length of residence was found to have a significant association with how 'at home' Landaffers felt, with more than 80% of those who had been in Landaff for more than 10 years feeling 'very' at home, as compared to 64% of newer residents. However, age group, gender and education were not significantly related to this measure of community attachment.

Despite Landaff's small size, most residents do not know the majority of adults in the town. Only about one-third (31%) of respondents said that they know most of the adults in town; none said they knew them all. At the same time, 25% claimed to know about half of the adults in town, 42% said they knew fewer than half, and 1% indicated that they knew none of the other adults. This is a reflection of both a declining 'density of acquaintanceship' (Freudenburg, 1986) resulting from the recent patterns of population growth, and the fact that most residents' daily activity patterns involving work, shopping and service acquisition extend outside of the local community. As one might expect, there was a statistically significant association between length of residence and the number of adults respondents said they knew in Landaff. Age, gender and educational level were not significantly associated with number of adults known by survey participants.

Fig. 6.5. Landaff, New Hampshire, 1942. Community church.

Survey participants were also asked about the numbers of their adult friends who were located both in Landaff and within a 1-hour drive of the town. Respondents had an average of ten friends in Landaff and 18 friends within an hour's drive. However, the majority said that most of their friends did not live in Landaff. Sixty per cent said that fewer than half of their friends lived in Landaff and 12% said that none of their friends lived there. There was a statistically significant relationship between length of residence and the number of friends residing in Landaff, with longer-term residents reporting more local friends than was the case for those who had lived in the community for less than 10 years. Despite the fact that the mean length of residency in Landaff was 20 years, it appears that many residents do not have close friends in town. One wonders how much social interaction there is between residents if most of their friends live an hour or more away. This is probably a change from the 1940s, when most or all of the farming families knew each other, and an indication of the social consequences of continued in- and out-migration over the span of the past half-century.

Nearly half of those responding to the community survey (49%) stated that they spend less than 1 hour per month taking part in organized group activities involving other members of the community. The other half spend anywhere from 1 to 4 hours per month to more than 10 hours per month participating in group community activities. When asked if the respondent or any member of their immediate household had participated in a community improvement project during the past year, the responses were roughly evenly divided – 49% had been involved while 48% had not. In bivariate analysis, no association between length of residence and the level of activity in community events was uncovered – newcomers were just as active, or inactive, as old-time Landaffers. Age, gender and educational level were also not significantly associated with number of hours spent per month by Landaffers on community activities. At the same time, there was a significant relationship between length of residence and the number of organized groups to which respondents belonged. The mean number of groups to which residents of less than 10 years belonged was just under one (0.7), while those who had resided in Landaff for more than 10 years reported belonging to an average of nearly two groups (1.7). Overall, Landaff residents claimed to be very interested in their community, but seemed to be divided fairly evenly between being active or not in community activities.

The education of Landaff's youth appears to be one of the most important and passionate subjects in town. In a survey question that asked about the importance of maintaining Landaff's autonomy and independent decision-making authority, the majority of respondents (56%) said that it was most important for Landaff to maintain its control over education.

At the time of MacLeish and Young's study (1940), there were three schools and school districts, each of which defined a neighbourhood. As Landaff's population declined, the more remote schools closed. Two schools (Scotland and Ireland) closed after the Second World War, leaving only the Blue School, a typical one-room schoolhouse. Today the Blue School is one of only two surviving one-room schoolhouses in New Hampshire (New Hampshire Department of Education, personal communication, 1997). The building remains a major landmark, both as a reminder of the past for many long-time residents and as a major connection to the community for recent in-migrants. According to Ploch, the school represents both the psychological and physical focus of the community. At the same time, if school consolidation trends continue, whether the Blue School will continue to serve this function is very questionable. Perhaps a forewarning of this trend is that while the Blue School used to provide education across all grade levels, by the late 1980s the school serviced only grades 1–5. By 1996, the school served just 14 pupils in grades 1–3, while older students from Landaff (grades 4–12) attended school in nearby Lisbon.

The symbolic meaning of the Blue School is evident in the high levels of engagement in school affairs reported by local residents. During a 1989 roundtable attended by Ploch, attendees reached consensus over the long-term use of the building. That is, they agreed that if the Blue School were no longer to be used for education purposes, the town should continue to own and maintain the building. The group offered several suggestions for a new use for the Blue School, including making it into a library, a health centre, or a community centre. Data from the 1993 community survey revealed that Landaff residents have continued in recent years to be fairly active in attending school board meetings, even if they did not have school-age children. More than one-quarter of survey respondents stated that they 'always' attended the annual school meeting. Fifty-nine per cent indicated that they knew all of the school board members, and 20% knew one or two of the three board members. In short, the local school remains an important icon of community identity. Thus, any attempt to close the Blue School, even if it made economic sense, would appear to represent a substantial threat to the community (Ploch, 1989, p. 46).

As noted earlier, at the time of the MacLeish and Young study the Grange was certainly the most important institution in the community. Currier and Clement (1966) noted that the Grange was recognized on several occasions by the State Grange for its ritualistic, home, community welfare work and literary programmes. In 1966, the Grange marked its 35th year of sponsoring the Old Home Day. However, during the mid-1980s, the Fire Department took over sponsorship of Landaff's most important community event. This transition reflects Ploch's (1989) remarks about the decreasing role of the Grange over the years, a change that has mirrored the decline of farming in the community. Nevertheless, in the 1993 survey, most Landaffers responded that the Grange still provided the most leadership in, and was the organization that had the most impact on, the community (Conner and Schmidt, 1996). The Fire Department and the Lisbon Lion's Club were also named, albeit less frequently, as pivotal community organizations.

Ploch saw the Fire Department as an important component of Landaff, both as a unit of the town government and as a source of social/psychological identification (Ploch, 1989, p. 32). Outside forces were the reason for its creation. Landaff had to support its own fire company in order to qualify for mutual fire aid from Lisbon. New male in-migrants who joined the volunteer fire department helped to offset the loss of out-migrating Landaffers. However, tensions grew as the newer members took a more active role and tried to modernize the department (Ploch, 1989, p. 59). The department, which had taken over the management of Old Home Day, instituted some changes to the event that were not well received by traditionalists. The most divisive event occurred when the department requested hourly compensation for time spent on fire calls. This request was denied at the Town Meeting. As a result, the fire chief and several younger members resigned (Ploch, 1989, p. 59).

The Role of Government

Landaff, considered a township by the Census of Governments, has been governed by a board of three selectmen for more than 200 years. The primary forum for decision making is the annual town meeting which has historically been both a social and political event. However, as Landaff has become more integrated into the larger world, and as the number of intergovernmental rules and regulations has increased, contemporary meetings have become primarily business-oriented. In his 1989 study of Landaff, Ploch noted that the number of issues to be discussed (each designated 'an article') at the town meeting (the order of discussion is determined by the articles' location on the town 'warrant') increased from an average of 10–11 in the 1941–1961 period to an average of 21–30 in the 1966–1986 period. In recent years, most of the issues addressed have focused on service provision and planning concerns, such as applying for grants to rehabilitate housing or to improve the local water and sewer system. Many of these types of issues are mandated or offered by higher levels of government, signifying the ways in which Landaff has been drawn into the larger society (Ploch, 1989, p. 31). Consequently, the role of the selectmen has become increasingly important in directing and coordinating Landaff's vertical linkages to external authorities and organizations.

The majority of survey respondents (39%) answered that it was through local government that they felt most involved in Landaff's decision-making process, though nearly one-third did not feel involved in any particular forum. Nearly two-thirds of respondents indicated that they considered it very (31%) or somewhat (36%) important to be involved in the community's collective decisions. Forty-one per cent felt that they were very or somewhat involved in current collective decision making. Nearly two-thirds (58%) of the respondents attended the annual town meeting always or most of the time. The majority of respondents (59%) knew all three selectmen.

Length of residence was significantly related to responses regarding the importance of being involved in Landaff's collective decisions. Newer residents, those living in Landaff less than 10 years, were most likely to feel that it was somewhat important to be involved, whereas the majority of longer-term residents felt that it was very important to be involved. Length of residence was also significantly related to how respondents felt about their involvement in Landaff's collective decision making. Only 3% of the newer residents felt very involved, compared to 24% of the longer-term residents. Not surprisingly, length of residence was also a significant indicator in knowing the town's selectmen, with 74% of the longer-term residents knowing all of the selectmen as compared to 42% of the newer residents.

Conclusions

Ploch's introduction to his 1989 report on Landaff noted that the original study conducted by MacLeish and Young (1942) was done at a time when Landaff was considered to be at the midway point of a stability–instability continuum. Despite substantial economic, demographic and social change over the ensuing five decades, Ploch concluded that by the late 1980s

Fig. 6.6. Landaff, New Hampshire, 1995. Its setting in the White Mountains has made Landaff an attractive spot for many families. The building of new and large homes is common.

Landaff was still a 'middle' community in terms of stability, although there were some indicators of increased instability. Landaff appears to be typical of many northern New England rural towns in its struggle to retain its Yankee values while attempting to modernize and survive economically. Urban residents flock to towns such as Landaff because they represent a way of life that perhaps only exists in memory or imagination.

The results derived from our restudy of Landaff, particularly those derived from the 1993 community survey, confirmed many of Ploch's observations regarding the existence of 'social tensions' between the old-time Yankee farm families and the newer populations of retirees and young professional families. We believe that such findings are representative of what has been occurring in many other small New England towns in recent years. For many such places the challenge of the 1990s has been to discover how communities can maintain their rural way of life while adapting to modern realities. A perceived need to pursue economic growth while simultaneously maintaining community identification emerges as a parallel concern and a common theme throughout the region.

The major issues in Landaff over the past 50 years recognized by inhabitants are: the change in land-use patterns and population trends; changing farming patterns from diversified family farms to just a few larger-scale dairy farms; dispersal of farm or community families as young adults out-migrate in search of employment; the in-migration of better-educated families with small children; the provision of schooling for Landaff's children; and the struggle to maintain the cultural traditions and lifeways of the past while creating a future made up of both oldtimers and newcomers. Interestingly, many newcomers seem to be more actively engaged in preserving the Yankee way of life than is the case among longer-term residents (Ploch, 1978, 1985). However, both groups seem to share a substantial affection for and commitment to the place. An overarching concern binding inhabitants is their interest in limiting unchecked development so that Landaff's primary asset – its rural setting – remains attractive to current and prospective residents. Despite the lack of local economic opportunity, Landaff appears to be a community of choice, not just birth.

It is clear that the organizational life of Landaff, even in the face of major changes that have altered both its physical and social landscape since mid-century, continues with vigour. The annual Old Home Day community celebration remains a signal highlight of the summer. Various chambers of local government and the local school board continue to meet

Fig. 6.7. Landaff, New Hampshire, 1995. A large farm family house. Many farms, however, are no longer associated with dairy enterprises, the former mainstay of the community.

on a regular basis and make decisions addressing local needs and concerns. Of course, centralization of the economy has weakened community identity, and, similarly, access to a broader choice of services and employment opportunities outside the community has created fewer occasions for local interaction. Yet the legacy for local decision making – stretching back to 1764 – continues to represent a fundamentally important local practice symbolizing great persistence and continuity.

Community and Social Well-being in Contemporary El Cerrito[1]

Richard S. Krannich and Clyde Eastman

Introduction

The small village of El Cerrito is located in San Miguel County in north-central New Mexico, approximately 65 miles south-east of Santa Fe and 45 miles south of Las Vegas. At the time of the initial Rural Life Study conducted by Olen Leonard and Charles Loomis, El Cerrito was assessed as having fallen between the extremes of the instability–stability continuum, with a high degree of cultural stability countered by growing instability with respect to economic opportunity and sustenance organization (Leonard and Loomis, 1941). El Cerrito has attracted substantial sociological attention since that initial study. No fewer than ten published works focusing on social conditions and changes in this small village have appeared during the six decades since the original research was conducted (e.g. see Loomis, 1941, 1959; Nostrand, 1982; Eastman and Krannich, 1995).

Looking down from the edge of the mesa above the village on to the buildings and farm fields clustered along a bend in the Pecos River, the physical features of modern-day El Cerrito would seem to imply a place virtually frozen in time (Fig. 7.1). At first glance, there seems little to suggest that the village has changed from the geographically and socially isolated island of traditional rural-Hispanic culture and community organization encountered 60 years ago by Leonard and Loomis. However, in the ensuing years El Cerrito has experienced dramatic economic, demographic and social upheavals. Yet unlike some other traditionally Hispanic villages in northern New Mexico that withered and eventually disappeared during this period, the community has endured.

© CAB *International* 2002. *Persistence and Change in Rural Communities* (eds A.E. Luloff and R.S. Krannich)

Fig. 7.1. El Cerrito, New Mexico, in 1941. An isolated and highly traditional Hispanic agricultural village, with buildings constructed of traditional adobe materials.

Although El Cerrito has exhibited a surprising capacity to persist in the face of seemingly great odds, both its short-term viability as a community and its long-term prospects for adapting to the pressures of ongoing social and economic change remain in doubt. The present research evaluates the characteristics of modern-day El Cerrito from the vantage point provided by Kenneth Wilkinson's (1986, 1991) theoretical perspective on the linkages between community and social well-being. First, we present an overview of the major dimensions of change that have affected El Cerrito over the six decades since it was initially studied by Leonard and Loomis. We then consider the extent to which contemporary El Cerrito exhibits the major defining traits associated with the sociological notions of 'community', and consider how those traits may be changing in response to both local and extra-local forces. Next, we address the question of whether the social context that exists in contemporary El Cerrito is consistent with the notions of social well-being identified by Wilkinson as deriving from local community social structures and processes. Finally, we offer some observations on factors that are likely to influence the degree of persistence and the patterns of change in community conditions in the coming years.

As with the initial studies conducted by Leonard and Loomis, our restudy of El Cerrito has been based primarily on qualitative field research methods. From 1991 to 1999 we engaged in an ongoing ethnographic investigation of social conditions and trends in the village. Each of us made numerous, relatively brief (1- or 2-day) field visits to the village to observe community activities and events and to conduct unstructured ethnographic interviews. Interviews of varying focuses and depth were conducted with nearly all of the current residents, as well as with a number of other former residents and intermittent visitors who continue to have property holdings, family ties or other linkages to the village. On several occasions student field workers also visited the village for brief periods and engaged in ethnographic data collection. In addition, structured face-to-face interviews were conducted with a total of 24 resident and non-resident adults who were present in the village on a Saturday in early April 1995, when more than the usual number of people were present because of a planned cleaning of the mud and debris from the irrigation ditch that supplies water to garden plots and pastures.[2]

The field notes and interview responses derived from these various forays into El Cerrito provide the foundation for the analysis and interpretations presented here. Although the data, for the most part, do not lend themselves to quantitative analysis or statistical summarization, they provide a rich qualitative understanding of the local social context. Major events and many personal anecdotes were described on numerous occasions to multiple observers, helping to reinforce the reliability of our findings. Information gleaned from interviews provides fairly comprehensive documentation of local activities and events that have occurred during the research period and, based on the recollections of those interviewed, of events occurring during the years preceding the initiation of this research.

Patterns of Change, 1940–1990

As noted above, El Cerrito appeared to be relatively stable when first studied by Leonard and Loomis in 1939–1940. Indeed, Warren (1978) characterized El Cerrito at that time as an example of a setting in which there was little evidence of the 'Great Change' that transformed the economic, social and organizational characteristics of many American communities by the mid-20th century. However, the apparent stability that was observed in 1940 was deceiving, for the village was in fact on the verge of economic and demographic collapse even as Leonard and Loomis conducted their study.

The seeds of El Cerrito's demise were sown four decades earlier, when the New Mexico Court of Private Land Claims rejected most of the San Miguel del Bado Land Grant claim in 1904. The court confirmed the legitimacy of land grant claims to some 118 acres of irrigated land and house lots in the village area of El Cerrito, but rejected all claims to the surrounding mesa lands, which had traditionally supported several large livestock herds maintained by village residents. Cerritoans continued to graze their animals on the mesa until 1916 and were able to homestead some land after that, but only two families were able to continue livestock production on anything approaching a commercial scale. Village men without livestock found it necessary either to work for large sheep ranching operations maintained by outside interests or, increasingly, to work outside of the village for a significant portion of each year, most often for the railroad or in agriculture.

In 1940 there were 26 households in the village (Leonard and Loomis, 1941). However, only 21 of these were continuously occupied, while five households were maintained by families that had moved away but still returned for occasional visits (Nostrand, 1982). The irrigated land in immediate proximity to the village provided a meagre subsistence for the 135 people living in the 21 resident households. Government relief programmes increasingly supplemented dwindling private employment as the Great Depression wore on. By the onset of the Second World War El Cerrito was in dire economic straits.

El Cerrito residents were thus primed to respond to the expanding employment opportunities that became available in 1942 and onward as a result of the Second World War industrial build-up and subsequent economic expansion in many of the nation's urban centres. By the time of the first restudy of the village, conducted by Loomis in 1956, 15 families of the 26 that were present as continuous or part-time residents in 1940 had moved away. Three families had disappeared as the result of deaths, and only eight resident families were left. This exodus from the village continued for another decade, until by the end of the 1960s only five elderly people comprising two households continued to live in the village (Nostrand, 1982).

Not surprisingly, a number of changes in social organization and community institutions accompanied the population decline. The village school, highlighted in the original Leonard and Loomis study as a key locus of community activity, was closed about 1950, and the building was sold and soon fell into ruins. Not only were the remaining schoolchildren bused to Villanueva, thus exposing them regularly to outside influences, but the village also lost its social centre. Because there was no longer any place to hold a dance or political meeting, an important part of community life withered. The Catholic church, which was also a major focal point of community life in 1940, stopped holding regular services when a priest was no longer available to travel to El Cerrito to serve the dwindling population of the village. The physical church structure continues to be as well-maintained today as it was 60 years earlier, but it is now used only rarely, and then primarily for funerals. Nostrand (1982) noted that a marriage ceremony held in the church in 1980 was the first wedding held in the village since 1958. Few, if any, weddings have been held there since.

However, certain other aspects of local life actually changed for the better during this same time period. The village was able to form a Mutual Domestic Water Consumer Association and drill its first water well. That was a major sanitary milestone, since the *acequia* (gravity-flow irrigation ditch) that diverts water from the Pecos River to irrigate gardens and fields

Fig. 7.2. El Cerrito, New Mexico, in 1993. While some buildings have crumbled over the years, most have been maintained or renovated, and the physical character of the village has remained largely intact.

surrounding the village was until then the only source of culinary water supplies. In 1952 the local Rural Electrical Administration extended a power line into El Cerrito and the village had electricity for the first time in its history. Not many years afterward a telephone line found its way into the village, and much later television satellite dishes also appeared as they became available for general public use. Thus, even as the population declined, modernization and linkages to the outside world were occurring.

As El Cerrito's population reached its nadir in the late 1960s many people were very pessimistic about the village's future. A number of houses and other structures crumbled into ruins as roofs were removed or fell away and adobe walls were exposed to the eroding effects of rain and snow. Property was offered for sale on very good terms to anyone with money, opening the way for the appearance of new residents who had no prior ties to the village. During the late 1960s and early 1970s much of northern New Mexico experienced a surge of rural population growth involving individuals linked to the counter-culture or 'hippie' movement of the time. During this period several outsiders, including a group of several Anglos, bought property in the village, and El Cerrito became ethnically integrated for the first time in its history. Several Anglos have been in residence at any one time continuously since then. More significantly, there was no resident at the end of the 1990s who had not lived outside the village for a substantial period. At this time, El Cerrito is still a spatially isolated refuge from the hustle and bustle of the outside world, but it is no longer populated by people who are unfamiliar with that world.

By 1980 there were 11 people residing in five continuously occupied households in El Cerrito, including two households occupied by Anglos who moved in during the early 1970s (Nostrand, 1982). By 1992 there were nine continuously occupied households, with two others frequently occupied by families dividing their time between El Cerrito and urban residences elsewhere in the region. Throughout the 1990s the population remained relatively stable, though some changes in composition have occurred, with some people moving in, others moving out, and still others oscillating between residence in the village and in outside locations. However, with a stable base of 25–30 residents and an increasingly diverse mix of children, adolescents, working-age adults and retirees, the village is more vibrant that at any time during the past 40 years.

A deliberate but steady renovation of the housing stock and some new constructions have accompanied the repopulation of El Cerrito. This reflects a renewed confidence and optimism in the village's future. Two long-established extended families, the Aragons and the Quintanas, now own a substantial proportion of all land in El Cerrito. Aragon and Quintana land holdings pass to the heirs in these families, but not outside them. Property owned by others who are not members of these two dominant families is also tightly held. Once again it would be extremely difficult for outsiders to purchase property in El Cerrito at anything other than very speculative prices.

In 1940 the residents of El Cerrito were heavily dependent on agricultural pursuits as the primary sources of income and subsistence. With

Fig. 7.3. El Cerrito, New Mexico, in 1941. Village children dipping water from the irrigation ditch. At this time, all cooking and drinking water was carried by hand from the ditch.

increased linkages to the outside world and with access to alternative sources of income, the degree of dependence on agriculture has waned considerably. We estimate that between one-third and one-half of the irrigable land is now actually irrigated. At various times in recent years two greenhouse operations produced gourmet vegetables and bedding plants for the farmers' markets and restaurant trade in Santa Fe and Las Vegas, though at present both are inactive. Scattered fruit trees around the village produce fruit if irrigated. Two or three households have substantial gardens. Most of the land that is irrigated, as elsewhere in *acequia* agriculture, is planted to hay and pasture, and several residents have haying equipment. While there are no longer sheep in the village, several families own a few cattle and fewer horses.

With the earlier demise of the school and the church as functional social institutions, the El Cerrito *Acequia* Association is now the only formal governmental structure that plays a significant and recurrent role in organizing community life and activities.[3] In this semiarid region *acequias* were dug at the founding of a new community, and have always been an integral part of community life and traditional Hispanic culture. Along with the church, the *acequia* has traditionally been the major focus of community organization in the villages and towns of the region (Crawford, 1988). Successful irrigation requires cooperation to maintain the system and to allocate the water. In the case of El Cerrito, the *acequia* continues to generate a level of solidarity that carries beyond agriculture to other community activities, even though at the end of the 1990s irrigated agriculture no longer contributed significantly to any household's income. Clearly some other explanation must be found for the amount of effort that goes into the annual *limpia* (ditch cleaning), particularly for those non-land holders who drive hundreds of miles to participate in a day of very hard physical labour when silt and debris must be shovelled by hand out of a nearly mile-long section of the ditch.

At the time of Leonard and Loomis's initial study of El Cerrito, the *funcion*, or celebration in honour of the patron saint, was the major festival in the village. As with other aspects of village social organization, the period of depopulation led to a gradual erosion of this traditional activity. In the old days the *funcion* lasted 2 days, but by the time of Loomis's 1956 restudy it was reduced to just 1 day, and it was no longer being held when our restudy began in the early 1990s. While the *funcion* has disappeared, the annual *limpia* had, by the late 1990s, taken on a definite festive air as a secular replacement for that traditional festival. Loomis (1958, p. 59) mentions that the *limpia* had formerly been 'the occasion of considerable festivity usually lasting about two days'. However, during the period of severe depopulation the annual cleaning and maintenance of the ditch became an almost impossible task, requiring as much as two or three 2-day weekends of hard work by the handful of able-bodied men who were available to participate. When we first attended a *limpia* in the spring of 1992 the num-

Fig. 7.4. El Cerrito, New Mexico, in 1993. The newly reconstructed irrigation diversion dam, funded with a cost-share grant obtained by village residents from the US Army Corps of Engineers, plays a crucial role in El Cerrito's persistence.

ber of workers available for the task had grown considerably, and the task was accomplished in the course of one 7-hour day. As the decade progressed, acquaintances of villagers from Santa Fe and Las Vegas began to join the work party in search of what some characterized as 'wholistic' experiences. Others came with tents, motor homes, food and drink to join in the festivities. In 1999 a record number of 65 workers completed the *limpia* by noon and then went to one household or another for a hearty lunch and a weekend of socializing. As discussed later in this chapter, the *limpia* is a key facet of the interactions and collective actions that reflect the persistence of community in El Cerrito (Fig. 7.5).

In the absence of the formal governmental and service organizations common to larger settlements, informal social interactions and relationships and the presence of informal leadership have been, and remain, crucial to the community's capacity to address collective needs. Leonard and Loomis made a major point that the lack of serious factional strife was an important criterion in the selection of El Cerrito for their study. A major conflict some decades before between two extended families (the Quintanas and the Manzanares) had resulted in legal expenses that exhausted the resources of both sides, but by the time Leonard and Loomis arrived the Manzanares family had quit the village and relative harmony had been restored. In more recent decades the Aragon family has risen to prominence as the largest land-owning family in the village. While members of the Aragon family definitely compete with the various Quintanas and occasionally there is overt antagonism, they have managed to cooperate effectively on major projects.

Earlier studies of El Cerrito have focused little attention on leadership in

Fig. 7.5. El Cerrito, New Mexico, in 1993. The annual *limpia*, or irrigation ditch cleaning, is crucial both to the maintenance of the irrigation system and the continuity of cultural traditions and social ties in El Cerrito.

various realms or on various projects. However, it became clear in our study that informal leadership involving both members of long-established Hispanic families and more recently arrived Anglo residents has played a key role in several major community accomplishments during the 1990s. A few of the more recently arrived residents brought very significant bureaucratic skills and experience with them, adding considerably to the community's capacity for responding to and addressing its needs and problems. For example, these informal leaders were able to write a proposal and shepherd it through the US Army Corps of Engineers to obtain about US$250,000 to build the new dam across the Pecos River. In another case a part-time resident who was born in the village and who was respected by everyone served as the *mayordomo* (ditch master) and provided very effective leadership that was critical to the maintenance and operation of the *acequia*. Whether this calibre of leadership will continue to emerge is an open question.

Leonard and Loomis emphasized the importance of the family in Latin American culture as a key dimension of community ties and interactions in El Cerrito. There was, and still is, much evidence to support that point. Yet it is also clear that the strong attachments to the community that are evident in El Cerrito go beyond immediate family ties to include the place where the family has lived. As documented by Nostrand (1992), the Hispanic culture of northern New Mexico places a strong emphasis on individual identification with, and attachment to, the native village and surrounding land areas. This is dramatically manifested during the *limpia*, but also at other times. Just as Alex Haley went back to Africa to find his roots, Quintanas, Aragons and others come back to El Cerrito to reconnect with their roots. Generations of the ancestors of several families are buried in the cemetery. In addition to his parents and wife, one current resident has four sons buried in El Cerrito. Even though he had established a successful career outside, he returned to the village as soon as he was financially able to do so. Different individuals phrase it somewhat differently, but they seem to be seeking a primal connection to their past and a respite from the demands and harried pace of the dominant urban culture. El Cerrito provides a sort of idyllic refuge, a retreat for people seeking to reaffirm an ethnic identity they feel slipping away. In the case of the younger generation, going to El Cerrito offers an opportunity to help maintain the *acequia*, the source of life-giving irrigation water, and to walk in their ancestors' footprints.

Elements of Community in Contemporary El Cerrito

Although there is a vast, and at times contradictory, sociological literature regarding the defining elements of community (see Hillery, 1955), Wilkinson (1986, p. 3) identifies three core domains that are both central to most definitions and key to an interactional approach to studying the community. The first involves *locality*, e.g. a territorially defined 'local ecology' which is the arena within which people engage in the activities of daily life and meet many, if not most, of their daily needs. The second element involves the presence of a *local society*, e.g. a localized organization of social life involving a 'full complement of the social structures through which a common life is organized' (Wilkinson, 1991, p. 27) in order to meet daily needs and express common interests (see Wilkinson, 1986, p. 3). The third element involves the presence of a *field of community actions* oriented toward locality-based needs and concerns (Wilkinson, 1986, p. 3 and 1991, pp. 18–39). Each of these major elements of the community provides a basis for evaluating the social context of contemporary El Cerrito and the changes that it has undergone over time.

Locality

Mixed conclusions can be drawn from a consideration of this dimension of the community as a criterion for evaluating the characteristics of contemporary El Cerrito. The village is very much defined, both in a physical sense and in a social and psychological sense, by its concentrated spatial pattern, its separation from other settlements, and the subsequent ways in which these spatial characteristics influence social activities and social structures.

Although for many village residents the majority of daily activities and interactions are concentrated within the very localized context of El Cerrito, several adult residents work at full-time jobs, attend school or have other obligations that require routine travel outside of the village (mostly in Las Vegas, about 45 miles away), and several resident children attend public schools in nearby Villanueva (about 8 miles by a rough, semi-improved road from El Cerrito). However, one gets the sense that those who routinely leave the village for work or to meet other daily needs maintain a degree of place identification that differs sharply from the sort of 'limited liability' described by some community scholars examining the place-based social linkages and obligations in other (primarily urban and suburban) residential settings (see Kasarda and Janowitz, 1974). There is little indication that social ties and obligations associated with these non-local activities serve to divert attention or attachment away from the village. Instead, there is a definite sense that most view El Cerrito as 'home' in both physical and symbolic ways. Interview data repeatedly reveal a tendency for residents to describe the village as a quiet, safe, peaceful and comfortable haven to which they gratefully return after their forays into the outside world.

Most Cerritoans derive their livelihoods either from some form of retirement or disability pension, are self-employed in activities centred in the village (e.g. growing organic produce for sale to regional markets and restaurants, woodworking crafts, etc.), or work intermittently in construction trades and other occasional labour occupations. For most residents daily activity patterns do not routinely extend beyond the local setting, even though their incomes are derived from extra-local sources. Those engaged in localized self-employment, such as the growing of garden vegetables, periodically travel to Las Vegas and/or Santa Fe to distribute their products, but may otherwise remain in the village for many days between these trips. Several residents devote considerable effort to large household gardens that provide an important component of their overall subsistence; a few also still maintain a few chickens, goats and other livestock, either on land holdings in the village or on the mesa that surrounds the village. These localized activities contribute to a spatial concentration of a considerable portion of residents' social interactions. In many cases it appears that activities in and around the household, and interactions with immediate family members, dominate the day-to-day lives of village residents.

Locality is also an important defining element for the linkages that bind many people who are not full-time residents to the village. In addition to those who are full-time residents, a number of property owners maintain a regular presence in El Cerrito but have their primary place of residence in Las Vegas, Los Alamos and other places in the surrounding region. Others, who live as far away as Pueblo, Colorado (230 miles by highway), also make frequent visits to work on their gardens and homes and to maintain social ties in the village. Interview records reveal that these part-time residents, and even intermittent visitors, tend in virtually all cases to express a strong connection to El Cerrito, and in many ways to consider it their 'home' in a social and symbolic sense. Many of those who maintain part-time households or unoccupied homes in El Cerrito express hopes and/or intentions to relocate to the village at some future date. Place-based ties to family, the land and cultural traditions contribute to a strong level of interest in and commitment to this place among both residents and those whose daily activity patterns are presently focused in other areas.

At the same time, El Cerrito is far from self-contained. With the exception of a locally provided culinary water system and connections to the outside world via commercial telephone and electric lines, residents must travel well beyond the local territory to obtain all public and commercial services. Interview responses indicated that residents rely primarily on service providers in Las Vegas, and to a lesser extent in Santa Fe, for groceries, medical and dental care, repair services, hardware supplies, appliances and other services. Schools, social welfare agencies, mail service and other government offices are accessed primarily through travel to Las Vegas or Villanueva.

In summary, the local territorial ecology of El Cerrito simultaneously concentrates most daily life in El Cerrito and forces residents to look beyond the local territory for employment, schooling and virtually all public and commercial facilities and services. Thus, a tension exists between a strong focus on interactions and sentiments linked to the locality, and activities and linkages involving the outside world. The unique social composition of the resident population allows a substantially greater concentration of daily activity in the village than is true in many rural communities, simply because most residents are not engaged in full-time occupations requiring regular travel to outside work locations. In addition, most residents appear to have made a conscious choice to forgo the conveniences of easy access to jobs and services and have adopted lifeways that allow them to focus much of their activity in the locality. There is considerable self-sufficiency in meeting daily needs. Gardens provide a substantial portion of the food requirements of several households. Backyard repair of cars, trucks and other mechanical equipment, building construction and home repairs are routine local activities pursued both individually and in cooperation with other residents who have the requisite skills. Trips for groceries and supplies, or to obtain commercial and governmental services, occur when necessary, but not frequently. Both by necessity and by choice, Cerritoans experience daily life patterns that are in most cases very much associated with place and locality.

Local society

While the day-to-day activities of most residents are concentrated in the locality, El Cerrito exhibits very few of the organizational and institutional features that are generally considered representative of a 'local society' through which residents could meet all daily needs and 'express all the major categories of the common interests of the people' (Wilkinson, 1986, p. 3). As was noted earlier, there are no formal organizational structures in the village other than the *Acequia* Association and the domestic Water Consumers Association. Even these operate largely via informal processes on a day-to-day basis. No formal local government structure exists, and there are no locally organized public service agencies. The church still stands and is kept in good repair, but regular services are no longer held. No clubs, voluntary organizations or other formal structures exist to link village residents in common activities and interests.

As Wilkinson (1991, p. 10) noted, inadequacies of social infrastructure can produce 'barriers to community interaction' and limit the capacity of local populations to respond to problems or pursue collective interests. This is undoubtedly true in El Cerrito, in part because the informal social structures that exist in the village do not always serve to break down such barriers. Some long-term residents suggest that the decline of the church, once the central focus of common activities and shared values (see Leonard and Loomis, 1941), has resulted in a decline in interaction and communication among villagers. In the absence of shared activities in local institutions and organizations, there is some indication of a tendency for villagers to keep to themselves, and for certain lines of social division to remain intact.

In addressing the question of social divisions it is important to note that relations between Hispanic and Anglo residents and landowners are characterized by a surprisingly high degree of accommodation across ethnic boundaries. However, there is some evidence of more substantial division within ethnic groups (see Eastman and Krannich, 1995). For example, long-standing tensions between members of the two traditionally dominant families in the village seldom give rise to overt conflict, but do appear to reinforce a tendency for much of the local interaction to occur among persons linked by kinship. Similarly, tensions persist among several of the more recent in-migrants over the dissolution of joint property ownership among the group of predominantly Anglo landowners who moved to the village in the 1970s and purchased property as 'tenants in common'. In the face of such tensions many people appear carefully to maintain a high degree of privacy, to focus their interactions and activities among members of their household and extended family members, and to engage in relatively limited informal socializing or neighbouring with others in the village.

The spatial concentration of El Cerrito and its very small population size contribute inevitably to an extremely high density of acquaintanceship (see Freudenburg, 1986). Residents exhibit a very high level of interpersonal familiarity, and it is virtually impossible for even a short-term visitor to remain anonymous. However, such familiarity does not necessary lead to the kind of emotional interpersonal bonding or 'communion' that is often presumed to characterize life in small rural places (see Wilkinson, 1991, p. 16). Conflicts and disagreements such as those evident in this village are both commonplace in most locales and by no means inconsistent with either the establishment of community sentiment and commitment or the pursuit of mutual interests. However, the absence of local institutions and organizations that might create opportunities for interaction and reinforce common interests and purposes serves as an important barrier to the emergence of certain community activities and processes in El Cerrito.

Community actions

The interactional approach to studying the community as set forth by Kaufman (1959) and elaborated upon by Wilkinson (1970b, 1979, 1986, 1991) focuses attention on a very specifically defined range of localized social relations that are indicative of the presence of the 'community field' from which 'community actions' emerge. Such interactions involve 'collective efforts to solve local problems and collective expressions of local identity and solidarity' (Wilkinson, 1986, p. 3). Community actions reflect 'shared interest in things local' (Wilkinson, 1991, p. 39), and purposive efforts to pursue and protect those shared interests.

In many ways, the extensiveness of community actions that occur in El Cerrito and the consequences of those actions comprise some of the most important reasons why the village has been able to persist, and even experience some degree of revitalization, over the past 20 years. Some of these actions have involved relatively focused, one-time efforts by a few capable leaders to accomplish some major objective, while others involve recurrent activities pursued by many individuals. In combination, these various community action episodes represent important keys to understanding how and why residents and property owners attach such strong meanings to the village, the land and the traditions associated with this place.

As we have noted elsewhere, several village residents who had previously lived in urban areas and worked in managerial and administrative positions have provided important leadership in efforts to pursue grants and other resources from state and federal agencies for improvement of local infrastructure. Through the talents and efforts of these individuals the village has successfully secured federal agency funding for a new dam for diverting irrigation water from the Pecos River, state funding to replace a washed-out bridge providing access to the shortest route to Las Vegas, and funding to establish the village's culinary water system (see Eastman and Krannich, 1995). These accomplishments are important in their own right, both as evidence of community activeness and leadership capacity and of the ability of local residents to engage outside authorities and pursue their common interests effectively.

However, several other examples of collective action in El Cerrito are perhaps of even greater social significance in terms of the breadth of participation that occurs and the extent to which such participation reveals the social bonds and meanings that characterize people's relationships with one another and with the community. For the most part, these revolve around the major work activities required to maintain the irrigation ditch that diverts water from the Pecos River for villagers' use on gardens and fields. Each spring (usually in late March or early April) residents, non-resident property owners and many others with far less immediate ties to the irrigated lands of the village converge on El Cerrito for the *limpia*, or ditch-cleaning. Over the course of most of a day and sometimes more, a crew of men (and, in recent years, a few women) use shovels to clear the previous year's accumulation of silt, weed growth and debris from the entire length of the ditch. Such participation is not entirely voluntary, as those property owners who have water shares are required by the *Acequia* Association to contribute a specified amount of labour to such endeavours (or to pay to hire labour in place of their participation), in proportion to the number of water shares that they own (see Crawford, 1988). However, in recent years the number of workers has been far in excess of what would be required of members in the *acequia* organization. Sons, grandsons and other relatives, who in many cases live hundreds of miles away and who may never have lived in El Cerrito, come to participate in this traditional community activity and the gathering of family members. In recent years the number of participants has been increased by the involvement of many non-residents whose only link to the village is their friendship with one or more residents and property owners.[4]

The social and symbolic significance of participation in the *limpia* is revealed both in the reasons that participants provide in explaining their involvement as well as the behaviours and interactions that occur during the activity. A number of those interviewed indicated that the activity represents an important tie to their family heritage and cultural traditions. Other reasons repeated with some frequency include a sense of responsibility to the village and to family members, the camaraderie that emerges among participants, and other social and recreational benefits. On several occasions participants have been observed instructing a young son, grandson or nephew on the finer points of clearing and shaping the ditch, clearly using the event as a way of passing on a sense of the value of hard work and the importance of taking responsibility and caring for the land, the water and the place. A sense of pride and accomplishment is clearly evident in the expressions of satisfaction, and even joy, among participants when this arduous task is progressing well, and especially when the work has been completed. The *limpia* provides an important opportunity for renewal and strengthening of social bonds, as evidenced by a great deal of visiting among friends and relatives and renewal of old acquaintances among those who live in other areas. Although the ditch cleaning is an arduous task, the spirit of the workers is often festive, so long as the work is proceeding well. Friendly bantering and joking prevail, and on occasion groups of workers will sing a song in Spanish, also indicative of the important link to cultural traditions that this activity represents.

In addition to this annual activity, a number of other less routine episodes provide evidence of the nature of community action in this small village. One especially noteworthy example involves local efforts to maintain the irrigation system and repair damage caused by high water and erosion. Over the years periodic high river flows have eroded away sections of the narrow embankment separating the irrigation ditch from the river. Walking along the strip of land separating the ditch from the river, one encounters long sections that have been built up and reinforced with dozens of very large, rectangular wire baskets (known as gabion baskets), each filled with hundreds of large rocks. When asked about this, villagers note with considerable pride that these were put in place several years ago through their own very substantial efforts. Because this section of the embankment is high and narrow, it is inaccessible to power equipment. As a result, all of the rocks were moved by wheelbarrow across the shallow river, and the baskets filled by hand, one rock at a time. One property owner and part-time resident noted the special significance of having each worker touch each of the rocks as they were passed along a human chain and placed into the embankments, noting that this activity reinforced both participants' sense of shared commitment and interdependence and the importance and power of community cooperation.

Similar episodes have occurred on a number of occasions, with the most recent example involving efforts undertaken in early 1996 to repair the diversion dam that had been damaged when flood waters cut around the end of the dam and eroded away much of the riverbank that had surrounded and supported parts of the diversion structure. Working with a small grant (US$2000) secured from the Corps of Engineers, villagers used river rocks, gabion baskets and concrete to accomplish a very substantial reconstruction of the embankment and diversion headworks. Such a project would likely have required tens of thousands of dollars and considerable bureaucratic delays if it had been pursued through normal agency procedures.

Overall, such events provide clear evidence of the kinds of social participation and actions that are central to an interactional approach to studying the community. Such activities are not evident at all times, but they are also not ephemeral. Established social structures and norms regarding responsibilities to the village, to family, to the land and to one's heritage provide a foundation from which such actions can emerge when the need arises. Due both to the social infrastructure provided by the formally organized *Acequia* Association and the cultural and symbolic importance of water, much of the action potential that is evident in El Cerrito is centred around maintenance of the irrigation system. However, the extent to which the social structures and processes that have evolved in response to addressing this particular locality-based common interest may be drawn upon in responding to other interests and problems is less certain.

Community and Social Well-being

In discussing the linkages between community and well-being, Wilkinson (1991) argued that consideration of the basic material needs of residents should be supplemented by an evaluation of ways in which community conditions may contribute to five major dimensions of social well-being. These well-being elements provide an additional basis for assessing the conditions observed in contemporary El Cerrito.

Material well-being

Wilkinson argued that the potential for social well-being to emerge as an outcome of community structures and processes hinges in part on the ability of local residents to secure the necessary levels of material well-being to meet sustenance needs; fulfilment of such 'lower order' needs is suggested as a necessary, but not sufficient prerequisite for social well-being to occur (Wilkinson, 1991, p. 72). When individual material needs are adequately met, there is increased potential for the emergence of social activities, processes and structures that can contribute to collective social well-being.

In El Cerrito the level of material well-being might seem a potential obstacle to residents' abilities to focus attention on collective community concerns. Only a handful of residents are employed in full-time jobs, and many must rely on relatively low fixed or intermittent sources of income. With the exception of vehicles driven by some non-resident visitors, nearly all of the cars and trucks that one sees in the village are at least 10 years old and often much older than that. Houses are solid but far from elaborate, furnishings functional but spartan, clothing plain and often well-worn.

Yet there is little to indicate that limited financial and material resources are a source of great concern or hardship. Compared to many other places, the costs for housing are low, and widespread use of wood for home heating helps to keep those costs low as well. The lack of social expectations and pressures to 'keep up' with others in terms of home and yard, automobiles, furnishing or clothing makes it easier to get by with limited financial resources. Also, the people of El Cerrito are extraordinarily self-reliant and adaptive. Drawing upon a combination of building and repair skills and individual resourcefulness, they construct and repair their own houses, maintain and rebuild vehicles and equipment, use scrap materials to create labour-saving mechanical devices such as powered maize grinders, and otherwise meet many needs that would force most people to seek the help of professional and commercial service providers. Perhaps more importantly, residents of the community have, in most instances, made an intentional, conscious choice to adopt relatively spartan lifestyles in exchange for the opportunity to live life at a more relaxed pace in a place that has strong emotional and symbolic importance for them.

Social well-being

In discussing the ways that the presence of community can contribute to social well-being, Wilkinson (1991, pp. 73–75) focused attention on five interrelated dimensions that provide a basis for evaluating the extent to which well-being is established. These include: (i) distributive justice; (ii) open communication; (iii) tolerance; (iv) collective action; and (v) communion.

Distributive justice

In discussing this dimension of well-being, Wilkinson focused attention on recognition that 'people are equally human', and the ways in which removal of inequalities would 'facilitate communication and encourage affirmative, accurate interpersonal responses' (Wilkinson, 1991, p. 73). Although the limited material resources of many village residents and the ethnic minority status of the Hispanic residents are important elements of inequities that operate at the broader societal level, there is no evidence of class differentiation within El Cerrito. Some residents and property owners certainly have more material resources than others. There are substantial differences in educational background, with some residents having no more than a few years of formal education and others holding graduate degrees. Some homes are in better repair, some vehicles more sound. Yet any differences in such external indicators of relative wealth and prestige are far less distinct than in most rural or urban settings, and do not seem to have much to do with social status in the local arena. More importantly, none of our data suggest inequities in the ways that people in the village relate to one another or their access to social activities and opportunities at the local level. Certainly there are some individuals who, by virtue of their personality or behaviour are viewed as less friendly, less helpful or less trustworthy. Similarly, some who possess certain kinds of experience or demeanor are looked to most often for leadership. However, any such distinctions seem to have very little to do with ascribed status, and much to do with status that is achieved through actions and capabilities demonstrated in local interactions and actions.

Open communication

A preponderance of informal, face-to-face social relations and a high density of acquaintanceship (Freudenburg, 1986) can contribute to 'efficiency of channels for transmitting information and resources among people' (Wilkinson, 1986, p. 73). In El Cerrito the absence of formal bureaucratic organization at the local level ensures that there are few institutional barriers to direct communication; the 'town hall meeting' character of decisions involving the *Acequia* Association provides a forum in which communication about the pursuit of collective tasks and needs can occur in a very direct manner.

At the same time, the latent tensions that characterize relations among some local families and individuals, and the limited social infrastructure available to channel interactions and communications, serve in some ways to restrict the completeness and authenticity of some exchanges. Avoidance of interactions that have the potential to create discomfort and perhaps conflict may limit capacity for collective actions. In the absence of shared involvement in regularly occurring local organizational activities or events, and the almost inevitable need to encounter and communicate with other participants, there is a tendency among some villagers to retreat into the privacy of their homes and their families. The structural and ecological barriers to communication are few – it is only a short walk to all but two houses in the village, and norms of reciprocity and egalitarianism make it unlikely that an effort to initiate conversation and pursue some need or concern would be rejected. However, among some residents the social barriers to such communication make it likely that interactions will occur primarily in the context of specifically focused issues and events, often those centred around activities of the *Acequia* Association.

Tolerance

The presence of certain social divisions and of some interpersonal tensions in El Cerrito does not imply an absence of tolerance. As is true anywhere, differences in ethnicity, age, family status, values, interests, backgrounds and behaviours influence the closeness of interpersonal relations and the frequency and nature of interactions. However, in El Cerrito there is clearly a great deal of accommodation of differences among the villagers, and a norm of mutual respect for the rights of others to be different.

One illustration of this is the accommodation across ethnic boundaries, with Anglo and Hispanic residents working closely and cooperatively in their efforts to pursue community needs and interests through work projects, interactions with outside authorities, etc. El Cerrito is unusual in that ethnic differences appear to be of little if any consequence with respect to friendship patterns, cooperation in community activities or the civility of day-to-day interactions. Also, there is a tendency to ignore or shrug off other differences. A resident may comment that some other individual is 'different' in some way – perhaps because of his or her tendency to be especially helpful or especially confrontational, or because of behaviours such as past or present use of some illegal substance. However, such differences are almost always simply noted in passing, and often accompanied by an observation like 'that's just the way he is'. Norms of accommodation and tolerance provide a key factor in the ability of this surprisingly heterogeneous village to both maintain a high degree of interpersonal civility and to mobilize in pursuit of collective interests.

Collective action

In Wilkinson's interactional perspective, 'collective actions representing the entire range of common locality-oriented interests' (Wilkinson, 1991, p. 74) comprise the key factor in determining both the emergence of the community and the presence of social well-being outcomes in a community context. Indeed, he essentially equated well-being and action, noting that

Fig. 7.6. El Cerrito, New Mexico, in 1941. At the time of the initial study by Leonard and Loomis, the village's Catholic church was one of the key village institutions and a major focus of community social life.

'social well-being entails people working together in pursuit of their common interests' (Wilkinson, 1991, p. 74).

As described above, locality-oriented interactions and collective actions revolving around the maintenance of the village's irrigation system comprise a central focus of the 'community field' (see Wilkinson, 1970b) that is evident in El Cerrito. Collective activities related to this common interest are not pursued on a daily basis, but both formal and informal social structures are in place to stimulate and organize such activities when needs arise. In addition to contributing directly to the material well-being of villagers, such activities help to build social relationships and give rise to a collective sense of accomplishment and a sense of common purpose and identity.

In contemporary El Cerrito this sense of common purpose is reinforced by a sense of common predicament (see Sherif, 1966) with respect to external authorities. Although the ability to secure funding and other resources from state and federal agencies has been an important part of El Cerrito's revitalization, village relations with such agencies are characterized by considerable tension, distrust and conflict. Ongoing efforts at the state level to adjudicate water rights, monitor levels of use and terminate the rights of those who have failed to maintain the required 'beneficial use' through continuous application of irrigation water are viewed as a direct threat to the material well-being of local residents and the sustainability of the village. Similarly, considerable hostility is focused on the US Army Corps of Engineers because of that agency's failure to either admit responsibility for or resolve design problems with the recently constructed dam that make operation of the irrigation system difficult and now appear to threaten the long-term integrity of the structure.

While the activities mentioned thus far are clearly indicative of the kind of collective action that Wilkinson associated with community well-being, it is important to make note of the almost exclusive focus of collective actions on issues related to the irrigation system. Given the limited social infrastructure and the lack of a critical mass of residents who share particular needs or concerns, some other potentially important matters of common interest to Cerritoans are less likely to be addressed through collective action. For example, several of those interviewed commented on the need for such things as improved fire protection, road improvements and activities for children and youths living in the village. However, there is no evidence of organized, collective efforts to address these and other needs, and such efforts would seem unlikely to occur given limitations in both human and fiscal resources.

Communion

This dimension of community well-being involves a 'consciousness of community and joyful response to the relationships that are realized' (Wilkinson, 1991, p. 74). As described previously, such 'communion' is clearly evident during the annual ditch-cleaning event. More than in any other aspect of local life, the *limpia* represents a context in which what, at face value, would simply appear to be a difficult, task-oriented activity actually assumes symbolic and emotional meanings that allow it to serve as an important catalyst for the emergence of common bonds among residents, property owners and even some outside participants. Such 'celebration of community' may be limited to this and perhaps other rare and special occasions, such as a wedding or the (now dormant) celebration of the village's patron saint, but the bonds that develop out of such experiences are clearly important in developing and sustaining collective action and other well-being dimensions.

Conclusions

According to the expectations generated by the theories of Wilkinson and other community scholars, El Cerrito represents a very 'incomplete' com-

Fig. 7.7. El Cerrito, New Mexico, in 1993. Although regular church services are no longer held in the village, the church building is still lovingly maintained by local residents.

munity, at least in terms of the nature of the local ecology and the limitations of local social infrastructure. The absence of locality-based organizations and structures to provide employment opportunities, and the lack of various facilities and services present in larger and less isolated places, force villagers to develop social and economic linkages that extend across a large territorial area. Nevertheless, the particular social composition of the resident population and the strength of cultural traditions and shared values help to create a surprising degree of 'communityness' here. Despite the very significant constraints of small numbers and limited resources, a surprising level of community action is evident, and such action has impressive consequences in terms of both task accomplishments and the ability to bridge interpersonal and family tensions, cultural barriers and other social differences to create a sense of collective purpose and a degree of communion. The kinds of well-being outcomes described by Wilkinson are very much in evidence, although problem areas remain unaddressed due to the relatively narrow focus and scope of local action episodes.

At the same time, it is important to recognize El Cerrito's vulnerability to forces that could easily cause an erosion of community cohesiveness and community well-being. Leadership is an important factor in the pursuit of community action (Wilkinson, 1970a), and in recent years it appears that El Cerrito has suffered a substantial erosion in local leadership capacity. Several of the respected informal leaders are aged and in declining health, and others have recently passed away. Some have found it necessary to move to more urban settings due to the physical rigours of life in the village and the need for more convenient access to medical services. During a recent *limpia*, several villagers commented on the death a few months before of one especially respected informal leader, noting that his loss was particularly significant because he had played an important role in pursuing cooperation and managing conflicts both within and between local families. Also, an Anglo couple who have played an important role in the past in pursuing development grants and in organizing construction work projects to provide employment for local men have become somewhat detached from village affairs, and now spend much of their time on Las Vegas business interests. The extent to which leadership roles will be assumed by others is not entirely clear. In the absence of respected leaders, the ability of the community to bridge social divisions and to secure continued involvement and participation in work activities among family members who do not reside in, or own property in, the village is uncertain.

Related to this is the gradual disappearance of villagers who possess a direct residential link to the village's past and its heritage. As older villagers die or move away, those who inherit homes and lands may be less strongly attracted to the village as a place of residence. On the one hand, some may feel less obligated to keep the property in the family, and more attracted to the financial gains that could result from selling. Should this occur, there would inevitably be some dilution of the cultural traditions and shared values that serve to reinforce the commonality of local interests. Such 'watering down' of traditional ties and of shared meanings associated with land, place and interactions has not yet occurred in El Cerrito, but such changes have had a major influence on the character of many other less-isolated villages in northern New Mexico. Already one vacant lot is being offered for sale (at a price that indicates the owner is fishing for a windfall gain). Locals also comment on the increased frequency with which land speculators and individuals interested in buying land drive through the village and ask about the availability of properties; several of those interviewed identified potential 'gentrification' as an important threat to the community. Recent patterns of rapid population growth in a number of outlying rural areas in New Mexico (New Mexico Bureau of Business and Economic Research, 1999) and the relative proximity of El Cerrito to Santa Fe suggest that opportunities to sell properties in the village will only increase for the foreseeable future. Thus the prospect of another depopulation appears unlikely, regardless of what may happen to the community's social organization.

There is a remarkable effort to socialize a new generation into village culture (especially noticeable during the *limpia*), and a strong adherence to tradition. In recent years some residents and landowners have initiated discussions of a proposal to preserve the traditional character of the community by restricting the movement of mobile homes or other non-traditional structures into the village. It is still next to impossible to buy property from a Quintana or Aragon, the two long-dominant village families. Their lands are transferred within each family, and there appear to be enough members in both families who are willing and able to take up whatever becomes available to preclude any pressure to sell to outsiders. Undoubtedly the desire in the dominant culture to escape urban pollution and congestion will continue to push people toward rural areas like El Cerrito. However, the very strong Hispanic attachment to community, place and cultural 'roots' is likely to result in only limited intrusions of outsiders into this village, at least in the short term.

The long-term sustainability of the community is also clearly threatened by potential difficulties in maintaining the irrigation system and village water rights. Major problems are likely if the structural problems with the diversion dam are not resolved, either through agency response or locally initiated repairs. If the dam were to collapse and the community could not secure government funds to repair or replace it, the ability to meet material needs through subsistence and commercial gardening would be lost. Much more significantly, a major focal point of community action would be eliminated. Similarly, any major restrictions on water use that could result from the processes of adjudication of water rights could undermine community sustainability and collective well-being. In either event, the common bonds and shared sense of purpose that contribute to collective action could give way to the hostility and bitterness associated with a 'dying community' rather than with 'community well-being' (see Padfield, 1980). El Cerrito clearly experienced such a period during the late 1960s and early 1970s which provided Anglo 'outsiders' the opportunity to move in. Although that door is now largely closed, the situation could change quickly if the 'glue' that binds the community into common purpose and action should dissolve.

In conclusion, contemporary El Cerrito represents an important example of how community can emerge and persist even under difficult circumstances, and how local initiative and community action can contribute in important ways to individual and collective well-being. At the same time, as is true of many other rural places, the challenges to sustaining this community are substantial. Geographic isolation, limited material and human resources, a renewed surge of migration into rural areas, and vulnerability to extra-local political decisions all threaten the long-term adaptive capacity of the community. In all likelihood, the changes that have occurred in the 60 years since Leonard and Loomis first examined El Cerrito will be matched by equally dramatic changes over the next half-century.

Notes

[1] This research was conducted as part of Regional Project NE-173, supported by the US Department of Agriculture. Additional support was provided by the Utah Agricultural Experiment Station (Project #UTA-00842) and the New Mexico Agricultural Experiment Station Project (Project #NM 1–5-27173).

[2] An unexpected heavy spring snow storm the night before the planned ditch cleaning resulted in a last-minute cancellation of the activity. As a result, fewer non-resident adults were present in the village on the day of the interviews than would have otherwise been the case, as many who had planned to drive in on the morning of the scheduled work day did not do so. At the same time, those who were present in the village were generally unable to leave due to muddy road conditions, and with unexpected free time available all of those contacted agreed to participate in the interviews.

[3] The Water Consumers Association is the other formal organization present in the village. The Water Consumers Association maintains the village well and culinary water system and collects water use fees from households, but does not hold meetings or promote other organized activities.

[4] In 1999, 65 individuals engaged in the ditch cleaning – at that time a record number of participants. Many more people were present in the village. Given the traditional gender roles that still characterize participation in such activities, many of the women were engaged in visiting and meal preparations while the ditch cleaning was in progress.

The Old Order Amish Community 50 Years Later[1]

A.E. Luloff, Jeffrey C. Bridger and Louis A. Ploch

Introduction

Of the six sites studied as part of the Rural Life Study series, the Old Order Amish of Lancaster County were selected to represent the most stable and enduring set of social relationships. This area of 185 square miles (Fig. 8.1), centreing around Leacock Township and including East Lampeter, West Earl, Upper Leacock, Earl, East Earl, Paradise and Salisbury Townships, is a largely agricultural region close to the metropolitan areas of Harrisburg and Philadelphia, Pennsylvania; Baltimore, Maryland; and Washington, DC.

When Walter Kollmorgen conducted his study in 1940, the Old Order Amish had lived in Lancaster County since the early 1700s, and while the Amish made concessions to the material culture of the 20th century, the Anabaptist farmers who migrated to Lancaster County from the Rhine Valley would still recognize the culture and lifestyle of their descendants. Following the biblical injunction to 'replenish the earth, and subdue it; and have dominion over the fish of the sea, and over the fowl of the air, and over every living thing that moveth upon the earth' (Gen. 1:28, quoted in Hostetler, 1980, p. 91), the majority of Amish males continued to pursue farming or a closely related occupation.

Over the past 50 years, Lancaster County has changed dramatically. In this chapter we document the most significant changes and describe how the Old Order Amish community studied by Kollmorgen has responded to forces just making themselves felt in 1940. As a framework for organizing our findings, we examine persistence and change in the major dimensions of community outlined by Warren (1972, p. 13). These dimensions, which are described in detail later, include: (i) local autonomy (which ranges from

© CAB International 2002. Persistence and Change in
Rural Communities (eds A.E. Luloff and R.S. Krannich)

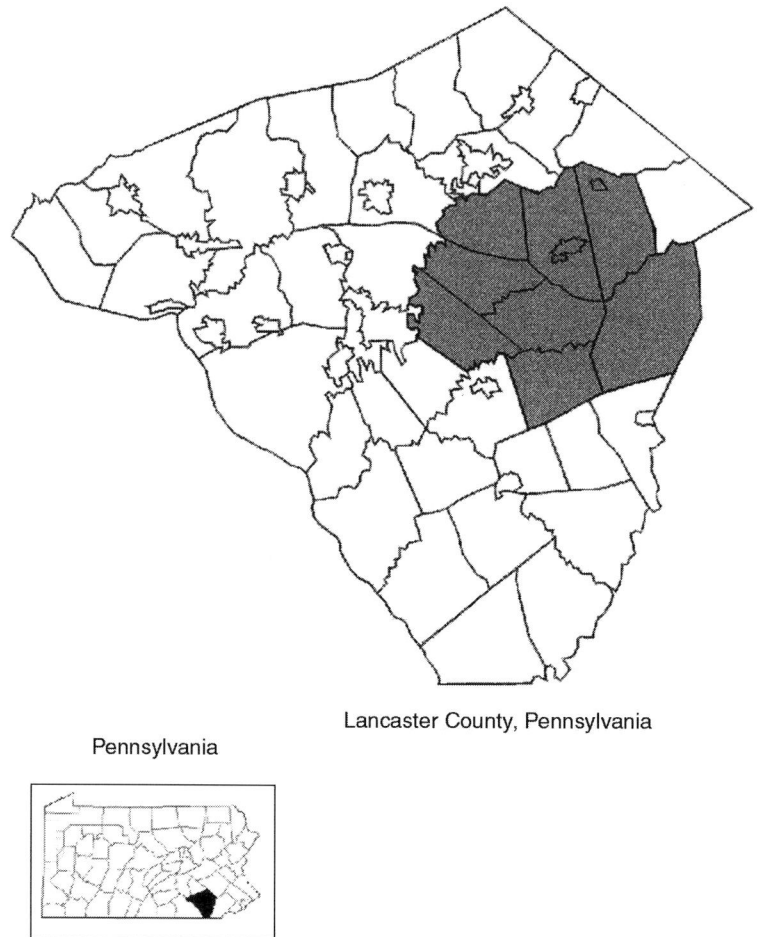

Fig. 8.1. Study site location.

independent to dependent); (ii) the community's horizontal pattern of networks and relationships (which ranges from strong to weak); (iii) coincidence of service areas (which can be centralized or dispersed); and (iv) psychological identification with locality (which ranges from strong to weak).

In order to chart the history of the Old Order Amish and Lancaster County from 1940 to 1990, we draw selectively on Kollmorgen's original study and compare this material with various studies that have been conducted over the past 40 years as well as data we collected through key informant interviews, county and local planning documents, and local newspaper coverage of events affecting the Old Order Amish. We begin with an overview of the demographic and land-use changes that have occurred within the study area and Lancaster County as a whole. We then move on to a discussion of the four dimensions of community life outlined above. The chapter concludes with some general suggestions for understanding persistence and change among the Old Order Amish.

Demographic and Land-use Changes

Between 1940 and 1990, the population of Lancaster County nearly doubled from 212,504 to 422,822 (Table 8.1). Within the eight-town study area, the aggregate population increased by 131% from 23,504 to 54,318. Each township except Earl more than doubled in size during this period, and East Lampeter more than tripled. In the face of competition for alternative land uses, the price of farmland increased substantially. Land-use pressure became even more intense after the Second World War with the proliferation of large-scale subdivisions of former agricultural tracts for housing, commercial and industrial uses. These conversions exacerbated the land market for those Amish wanting to continue in farming and contributed to the beginning of the search for other areas – both in-state and out-of-state – for the Plain Folk to farm.

As Lancaster County has grown, it has also become more diverse (see the Appendix to this chapter). Latinos, who migrated to the area in the 1950s to work as farm labourers, now account for 3.7% of the population. Even within the study area, which was virtually all White in 1940, all municipalities except Paradise now contain a mixture of racial groups. Similarly, the number of people over 65 has grown from 3.5% to about 11% between 1940 and 1990. Obviously, part of this increase can be attributed

Table 8.1. Population trends for study site townships, Lancaster City and Lancaster County, 1940–1990.

Areas	1940	1990	Change
Earl	2,875	5,515	91.83
East Earl	2,638	5,491	108.15
East Lampeter	3,847	11,999	211.91
Leacock	2,329	4,668	100.43
Paradise	2,574	4,430	72.11
Salisbury	3,579	8,527	138.25
Upper Leacock	2,964	7,254	144.74
West Earl	2,663	6,434	141.61
Study site total	23,469	54,318	131.45
Lancaster City	61,345	55,551	−9.44
Lancaster County	212,504	422,822	98.97

to the longer life span of contemporary Americans. However, it can also be tied to the growing number of retirees who have settled in the area. In recent years, Lancaster has become a popular retirement destination, and several retirement villages and rest homes have been developed to accommodate this segment of the population (Stranahan, 1993; see the Appendix for more complete details of changes in sociodemographic characteristics).

The land-use changes that have accompanied population growth are striking. As the figures in Table 8.2 illustrate, the amount of land categorized as developed more than tripled between 1947 and 1990, moving from 5868 acres to 18,294 acres. The majority of this land came from agriculture (10,969 acres) with an additional 1448 acres moving to developed from former forested land. In addition, the amount of land in the idle category increased by more than 500%, with 436 acres moving from agriculture and 52 acres from forested.[2]

Most of the conversion of agricultural land to developed land occurred along several highways which cut through the western portion of the study site. Since the 1950s, housing developments, a variety of commercial establishments and industries of all sizes have sprung up in this area. Highway construction also helped open the area to tourism; the number of visitors has grown from 2 million in 1965 to over 5 million by the late 1990s. Shopping outlets and a wide variety of amusement and historic sites based on Amish culture have proliferated in the study area, contributing a substantial portion of the US$450 million generated by the tourist industry.

Local Autonomy

At first glance, the Old Order Amish, with their distinctive lifestyle and well-known avoidance of the larger society, appear to have escaped the

Table 8.2. Lancaster County study site area land use, 1947 versus 1990 (acres).

1990 Land uses: 1947 Land uses:	Developed	Agriculture	Forest	Idle	Water	Total (1947)
Developed	5,859	2	1	6	0	5,868
Agriculture	10,969	87,981	445	436	0	99,831
Forest	1,448	782	9,782	52	17	12,081
Idle	18	0	5	89	3	115
Water	0	0	0	0	809	809
Total (1990)	18,294	88,765	10,233	583	829	118,704

forces of modernization that have overwhelmed most communities (Vidich and Bensman, 1968; Warren, 1972; Gallaher, 1980; Allen and Dillman, 1994). In the early days of the settlement, when land was plentiful and farming was the only fully acceptable occupation for Amish men, this may have been a fair characterization. The rich limestone valleys of Lancaster County, which locals describe collectively as the 'Garden Spot of the World', contain some of the most productive non-irrigated farmland in the nation. Here, the Amish were able to develop a thriving community based largely on agriculture.

Although the religious injunction to 'keep the garden, protecting it from harm through the use ... of labor and oversight' (Hostetler, 1980, p. 117) contributed to the effectiveness and productivity of Amish farmers, high land values also played a role. By 1938, with the growth of the Amish and Mennonite communities and constant improvements in their farm buildings, land prices were becoming prohibitive. In fact, while comparable land in other parts of the county was selling for US$75–100 per acre, land in the centre of the Old Order Amish and Old Order Mennonite communities was regularly selling for US$300–400 per acre (Kollmorgen, 1942, p. 27). In order to provide the land necessary for their children to remain in farming, the Amish regularly subdivided their properties, contributing to small farm size. This situation was a major factor in the growing importance of tobacco, potatoes, tomatoes, peas, livestock and other products that could generate much needed cash. Commercial agriculture was becoming a necessary means of maintaining a viable farming community: 'Commercial agriculture was not carried on for its own sake, but as a means whereby the farmer sustained himself and his children on the land. Farming was not practiced to make money, but money was made to support the farm' (Kollmorgen, 1942, p. 46).

Commercial agriculture alone, however, could not generate the capital required to purchase the amount of farmland needed for a constantly expanding population. By the early 1940s, young Amishmen were looking beyond the community for loans:

> During the early colonial period, farmers here apparently never borrowed money from non-Amish people or agencies. In time, they borrowed money from banks, on personal loans. In recent years they have also began to borrow from the production credit association. ... [A] number of non-Amish informants in Lancaster County have been approached, in recent years, by young Amishmen for personal loans that would enable them to begin farming.
> (Kollmorgen, 1942, p. 52)

Despite their growing reliance on cash crops and livestock, the Amish in 1940 still raised a large portion of their food. Every farm had a vegetable garden which was the province of the wife/mother. She was expected to can, dry and preserve most of the family's food supply (Kollmorgen, 1942, p. 47). Among the younger generation, however, the cost, time and effort

Fig. 8.2. Old Order Amish, Pennsylvania, 1942. Downtown Lancaster farmers' market in 1942.

required to preserve most of the family's food supply was being questioned and many housewives were beginning to purchase items that previously would have been produced on the farm. As Kollmorgen (1942, p. 47) noted:

> Self-sufficiency is in retreat. In the summer of 1939, five commercial bakers had bread routes running through the Amish country to serve an increasing number of patrons. Grocery trucks and meat trucks pass through the community to serve numerous customers. During apple and peach seasons, fruit vendors come up to the Amish houses confident of making sales. The grocer is selling the Amish farmers more and more canned vegetables.

The declines in self-sufficiency and autonomy that were becoming apparent in 1940 have continued. Although the Amish are still autonomous in comparison with most communities, they interact frequently with the 'English' and are increasingly linked to the larger society. Economic ties, in particular, have become more numerous as off-farm employment has increased substantially with the growth of tourism and escalating land prices. The editor of an Amish directory summarized some of the most significant changes that have occurred over the past 50 years:

> When our fathers implanted our church in Lancaster County it was their will, even a part of their faith, to live on a farm and raise their family, to live together and work together. The immigrant Stoltzfus family was the backbone of our community here. They had an unwritten motto to live together, worship together, stay together, and die together. That cohesiveness is one fundamental that built our churches to what they are today and was certainly an instrument

in building our county to the state of being named 'The Garden Spot of the World'. Until 1940 very few Amish heads [of families] were employed in off-farm jobs. By 1960 this practice rose and we began to realize the squeeze of available farmland diminishing and a fast-growing population. The conservative families that could not farm took to building and operating repair shops – to repair farm implements, harnesses, buggy shops (and more) which were farm related, needed by the commonwealth of the church and relieved the growing pressure.

(Anonymous, quoted in Testa, 1992, p. 59)

According to a recent study, nearly half of Amish males are now employed in off-farm occupations, including masonry, carpentry, cabinet-making and variety of entrepreneurial ventures (Smith *et al.*, 1994). The production of wood products is the largest area of activity, followed by metal products, general retailing and agricultural products. Most Amish-owned businesses are relatively new, with 61% established since 1980 (Kraybill, 1989).

The development of commercialized agriculture, off-farm employment and reliance on outside credit have enabled new generations of Amish to remain in Lancaster County, and there are considerably more Amish today than 50 years ago (Fig. 8.3). In 1940, there were 18 church districts containing approximately 3000 people. Now, there are over 16,000 Amish residing in 95 church districts (Kraybill, 1989).

Despite their population growth and material success, concerns have arisen about the long-term viability of the Amish community. Some church leaders, especially those of a more conservative bent, view material prosperity as a test of faith and fear that increased interaction with the outside

Fig. 8.3. Old Order Amish, Pennsylvania, 1995. Lancaster farmers' market today.

world will engulf the Amish community. The following editorial, which appeared in one of Lancaster's Amish publications, highlights the challenges posed by growth and development:

> Twenty years ago there were columns written in the newspaper that the tourist industry is reaping great profits at the expense of the Amish. In the last ten years this trend has reversed: *Many 'plain people' are now profiting from the tourist.* It is surprising that so little is talked about it among our people. What we hear comes from liberal groups saying things as: 'Oh we never had it so good. Our church districts are growing at an all-time high rate, we are prospering. What a blessing!' The more conservative or concerned people will say: 'We are living in the latter days of time,' which is definitely the whole truth.
> (Anonymous, quoted in Testa, 1992, p. 60)

The editorial concludes by recounting the biblical tale of Sodom and warning that the Amish will face a similar fate unless they obey more carefully the admonishment to 'be not of the world'. In the view of the more conservative Amish, efforts to cope with the rapid pace of change in Lancaster threaten to blur the lines separating the Amish from the 'English'. They worry that such adaptations as off-farm employment, participation in the tourist industry and the increased use of telephones and battery-powered electrical tools for business purposes, if not controlled, will lead eventually to assimilation.

The increasing diversification of Amish society has also raised troubling questions about the long-term stability of the self-sufficient mutual aid system that provides assistance to victims of disaster, the sick, the elderly and other less fortunate members of the community. Traditionally, the Amish have strictly avoided all ties with governmental relief agencies and opposed most investments in outside financial organizations. As Kollmorgen (1942, p. 22) noted:

> The Amish take care of their own poor, and none of them is now on relief. Members in distress still receive aid from fellow members, particularly if reverses result from 'acts of God'. Investments in outside organizations are discouraged and earned capital is kept in the community through loans to children or other members.

The Amish first became concerned about the future of the mutual aid system when Social Security was expanded to cover self-employed farmers in 1955. Amish reaction to this change was predictable. They argued that they had no need for such a programme because their support system already provided for the elderly and infirm. They also noted that participation in a state-sponsored welfare system would violate their constitutional guarantee of freedom of religion. Finally, participation in Social Security would demonstrate a lack of faith in God to provide for His children (Ferrara, 1993, p. 129).

Amish resistance to Social Security took several forms, ranging from refusal to pay taxes, to extensive lobbying, to lawsuits. Finally, an exemption to Social Security was attached to the Medicare bill passed by Congress in 1965. Under this legislation, self-employed workers were qualified for an exemption if they were 'members of a recognized religious sect with established tenets opposed to accepting benefits of any private or public retirement plan of life, disability, or health insurance' (Ferrara, 1993, p. 137). Those seeking an exemption needed to file before they were entitled to receive benefits and they also had to agree to waive all rights to future benefits for themselves and their dependents (Ferrara, 1993, p. 137).

After the exemption was passed, the Amish turned their attention to expanding it to the growing number of workers who were not self-employed. In their view, the same arguments applied to all Amish regardless of their employment status. Lawmakers, however, were not as receptive to this line of reasoning. When lobbying failed, an Amish employer who refused to pay the employer's share of Social Security tax sued the IRS for an injunction against collection efforts and for a refund of a small amount of Social Security tax that had already been paid. The federal district court ruled that forcing the employer to pay the tax would violate his right to the free exercise of religion. On appeal, however, the Supreme Court in 1982 rejected the plaintiff's argument, holding that an expanded exemption would open the door to claims by other groups seeking exemptions to a wide variety of taxes on the basis of religious beliefs (Ferrara, 1993, p. 139). After this defeat, the Amish redoubled their legislative efforts and were granted an expanded exemption in 1988 which applied to both Amish employers and employees. Currently, the only Amish who must pay Social Security tax are those who work for non-Amish employers.

In recent years, the proliferation of local regulations has emerged as one of the most significant threats to community autonomy. For most of their history in Lancaster County, the Amish have been able to farm and develop their land with very little governmental interference. However, as population growth, development pressure and concern over agricultural pollution and its impact on the Chesapeake Bay mounted, local governments exercised greater control over land use. Numerous regulations now affect farming practices, construction, zoning, subdivision and development and sewage management (Place, 1993). The Amish frequently find themselves in direct contact with public officials, marking a decisive shift in their relationship with the state (Place, 1993).

Farming practices in particular have come under heavy scrutiny as population growth has intensified over the past two decades. Excessive applications of manure, grazing cows along unfenced streams, and other traditional aspects of agriculture are now viewed as serious threats to groundwater and the wells that supply drinking water to many homes in Lancaster County. Moreover, as efforts to clean up the Chesapeake Bay began to focus on non-source point pollution, agriculture was identified as a major culprit. Lancaster County, which is the most intensively farmed area along the Susquehanna River, was singled out as one of the largest contributors to the excess nutrient levels that fuel eutrophication of the Bay (Stranahan, 1993, p. 221).

Historically, the Amish have applied large amounts of manure to their fields – a practice which contributed to higher than average yields (Kollmorgen, 1942, p. 16). When the density of farms was lower, heavy applications of manure did not represent a serious pollution threat. But as the Amish have simultaneously subdivided their farms and intensified livestock operations, their contribution to ground and surface water pollution has increased dramatically.[3] Agricultural runoff is exacerbated by the Amish preference for mould-board ploughing (which leaves no crop residue on the fields), the practice of spreading manure on frozen ground, reluctance to plant winter ground covers that hold soil in place, the planting of all available land to maximize crop yields, and refusal to participate in government-funded soil conservation programmes (Place, 1993; Stranahan, 1993).

In the face of growing public concern over water quality, state officials now enforce environmental laws that were rarely applied to the Amish in the past. In 1989, for instance, two Amish farmers were fined by the Pennsylvania Fish and Boat Commission under the Fish and Boat Code,[4] for allowing water to run off barnyards and pastures into gullies that presumably drained into streams. These incidents were particularly troublesome to the Amish because the farmers who were fined were following what they believed to be responsible practices (Place, 1993, p. 194).

Shortly after this incident, non-Amish friends organized a meeting between the Amish and government officials. The meeting provided an opportunity to learn more about the law and possible future regulations. However, although the meeting quelled some initial fears, additional environmental regulations have since been imposed. In 1993, Pennsylvania became the only state in the Chesapeake Bay watershed to make nutrient controls mandatory. The Nutrient Management Act required all farms over 10 acres to develop a nutrient management plan and institute best management practices within 5 years. In conjunction with a nutrient management specialist certified by the Pennsylvania Department of Agriculture, each farm needed to develop detailed plans that address nutrient availability and crop requirements, nutrient application rates and procedures, manure management procedures, soil runoff and identification of alternative manure storage areas to be used when normal manure handling procedures could not be followed.

The Amish have not been silent about what they see as excessive and unnecessary governmental intrusion. In the early 1990s, they attended a series of local hearings about the proposed Nutrient Management Act, signed petitions against the bill, and attended a state hearing during which the petitions were presented to officials (Place, 1993). The ongoing dialogue with state and local officials has lessened fears about governmental regulation, and the Amish are showing a greater willingness to cooperate with officials to protect the environment (Place, 1993, p. 198).

Nevertheless, many Amish remain concerned that state interference will eventually force farmers out of business.

Similar controls have been extended to other aspects of land use over the past few decades. As was the case with farming, the Amish have traditionally developed their land with little or no interference from the state. When an Amishman wanted to construct a *grossdaadi* haus (an addition to the main house built for elderly parents) or erect new farm buildings, for instance, he typically proceeded 'without concern for zoning ordinances, building permits, land development plans, hydrogeological studies, land surveys, storm water management plans, or sewage permits' (Place, 1993, p. 202). In most instances, local officials were willing to 'look the other way' out of respect for the Amish belief in self-sufficiency and separation from the world.

When construction and population growth skyrocketed in the 1980s, however, land use and development came to dominate county and township politics. Newcomers, in particular, worked hard to create policies that reflected the bucolic vision of rurality that had drawn many of them to the county – a vision firmly rooted in the Amish experience and their commitment to stewardship. According to the *Intelligencer Journal* (1985, p. 10), one of Lancaster's largest newspapers, the county is defined by a unique relationship between people and the land: 'Lancaster County farms – rich in soil, small and highly diversified – are operated by families with a long respect for the land and the highest standards of stewardship. The rich farmland is treated with respect and handed down from generation to generation.' Opponents of growth seized on this version of Lancaster's history to argue that greedy developers and corrupt local officials were undermining this delicate balance.

By the late 1980s, this perspective was gaining widespread legitimacy and local officials felt pressure to provide tangible evidence that they were in step with public sentiment by enacting new ordinances and more strictly enforcing existing regulations. Ironically, the Amish, who were deployed as such a potent rhetorical weapon in growth control debates, were now feeling the unintended sting of policies ostensibly devised to protect them. In order to maintain an appearance of impartiality, local officials now enforce land-use and subdivision ordinances more uniformly and are hesitant to make exceptions for anyone, including the Amish. Unfamiliarity with the many levels of bureaucracy involved in obtaining approval for land development has made the process costly and confusing. More importantly, some technological requirements are difficult to accept on religious grounds, leading some Amish to defy regulations they feel are particularly offensive.

The most controversial case in recent years involved David Fisher, a very conservative Amishman who constructed an illegal outhouse and chose to go to jail rather than comply with local sewage regulations. In a letter to a local sewage enforcement officer, Fisher outlined his objections to a modern sewage disposal system:

> You see, our whole reason for living what you may call a deprived life, what some call a simple, God-fearing life, is to be obedient to the Bible in being outside the entanglements of the world system. Those who might choose to live in the luxury of city life also take upon themselves the responsibility to live the life of building permits, social security systems, drivers licenses, etc. Those who avoid these as sinful are led to accept farm life, more specifically, the subsistence of self-sufficient farm life, as a distinct way of life.
>
> (Fisher, 1989, quoted in Testa, 1992, p. 126)

After serving 16 days of a 32-day sentence for failure to install a proper septic system, David Fisher was released from the Lancaster County Jail. Shortly thereafter, he and his family left the county and settled in western Pennsylvania.

David Fisher's experience was atypical. Although there are inevitable conflicts with the state, they are usually resolved in ways that preserve community autonomy surrounding fundamental religious issues. In fact, the Amish are increasingly taking a proactive approach that seeks to resolve confrontations before they escalate to the point where it is impossible to find middle ground. When new zoning restrictions limited industries on farms, for instance, a group of Amishmen hired a planning consultant to draft a zoning code that was acceptable to both township supervisors and the Amish.

Despite the gradual loss of autonomy in many areas of life, the Amish have managed to maintain tight control over the education of their children. By limiting education to the first eight grades, the Amish not only gain an important source of labour, but also restrict their children's exposure to foreign ideas during the critical teenage years, increasing the likelihood that children will join the church when they become adults. During the first few decades of the 20th century this practice went unchallenged; Amish children attended traditional one-room country schools with their 'English' peers through the eighth grade. Parents exerted considerable influence over the content and the structure of the educational experience, keeping a close eye on the curriculum, serving on school boards, attending parents' meetings, raising money for school equipment, donating labour and attending school functions. They also helped to select teachers and elect school board members (Hostetler, 1990, p. 259). By the time Walter Kollmorgen arrived in Lancaster, this situation was beginning to change.

In 1937, a large, federally funded, consolidated school was planned for Leacock Township – the heart of the Amish community. The Amish were initially divided over how they should respond to this development. One faction argued that Amish children should be withdrawn from public school. Others were afraid to disobey the law (Hostetler, 1990, p. 261). An 'Old Order Amish School Committee' was formed and, on the advice of a lawyer, they sought to prevent the construction of the school through legal channels. A petition, signed by 3000 people, asked state officials to allow Amish children to attend school for 8 months of the year in one-room

Fig. 8.4. Old Order Amish, Pennsylvania, 1942. Little Red Schoolhouse near Churchtown.

schools and exempt them from compulsory education beyond the eighth grade. State officials rejected the petition and the school was built.

At the same time, changes in Pennsylvania law required children to attend school through the age of 17. Those who were needed for farm work were allowed to apply for a permit that excused them from school at age 15. For the Amish this still meant that their children would have to attend at least a year of high school. As a means of circumventing this regulation, many Amish children intentionally repeated the eighth grade. When schools announced that they would no longer tolerate this practice, conflict was inevitable. Parents refused to send their children to high school and were brought to court and fined. When they refused to pay their fines, they were jailed (Hostetler, 1990, p. 263).

Finally, in 1955, after many more confrontations with state and local school officials, the Amish and the Commonwealth of Pennsylvania reached a compromise. Governor George Leader arranged for a reinterpretation of the school code which legitimized the Amish vocational school. Under this plan, Amish children could leave school at the age of 14 to perform farm and household chores under parental supervision. They were required to keep a daily journal of their activities and attend 3 hours of instruction per week. The vocational schools were required to teach certain subjects and record attendance, but they did not require a certified teacher (Hostetler, 1990, p. 263). The Amish School Committee continues to coordinate and monitor educational activities and, since 1957, Amish communities throughout Pennsylvania have held annual meetings to set guidelines and standards for Amish schools (Meyers, 1993, p. 92) (Fig. 8.5).

Fig. 8.5. Old Order Amish, Pennsylvania, 1995. Typical schoolhouse today; structure and style remain the same.

Horizontal Pattern

In both scholarly and popular accounts, the Old Order Amish of Lancaster County are described as the quintessential close-knit community – the kind of place where lives and homes are created through hard work and cooperation, where 'everyone contributes toward combating disasters, building civic institutions, mourning losses, and celebrating joys, ... where mutual dependence mutes individualistic impulses toward unrestrained self interest, ... where friendliness and mutual support are a way of life' (Morris and Hess, 1975; Jakle, 1982, quoted in Hibbard and Davis, 1986, p. 420). In short, the Old Order Amish are held up as a classic example of a community with an exceptionally strong horizontal pattern, which Warren (1972, p. 13) defined as 'the structural and functional relation of the various local units (individuals and social systems) to each other'.

In large part, this description of the Amish is rooted in their experiences in Lancaster County. When William Penn granted land to the Amish, they made clear their intention to avoid participation in worldly affairs: 'We do not attend elections, we enter not your courts of Justice, we hold no office either civil or military' (Kollmorgen, 1942, p. 75). The Amish belief in separation from the world is based on a literal reading of the Bible. Two passages exemplify the central message that the Amish take from the Bible (Hostetler, 1990, p. 75). The first is the injunction to: 'Be not conformed to this world, but be ye transformed by the renewing of your mind that ye may prove what is good and acceptable and perfect will of

God' (Rom. 12:2). This doctrine justifies Amish existence as a 'peculiar people' through language, dress, behaviour and restrictions on the adoption of certain technologies, among other things. The second passage admonishes the Amish to avoid entanglements with unbelievers and sinners: 'Be ye not unequally yoked together with unbelievers; for what fellowship hath righteousness with unrighteousness? What communion hath light with darkness?' (2 Cor. 6:14). To the Amish, all outsiders are unbelievers, and intimate relationships are prohibited.

Although the belief in separation from the world set the stage for a strong horizontal pattern, other factors also were important. First, the universality of farming and closely related vocations created a common universe of discourse and a common occupational bond (Warren, 1972). In Hawley's (1950) terms, the Amish in 1940 could best be described as a categoric group, 'an association of functionally homogeneous individuals' (Warren, 1972, p. 57). In contrast to more differentiated communities, where neighbours might have little more than physical proximity in common, the Amish were united by the knowledge, problems and demands arising from their common reliance on agriculture.

The unity created by occupational homogeneity was strengthened by ethnic homogeneity and intermarriage, which created a dense network of kin ties:

> The community is completely uniform in ethnic composition, for all ancestors of these people came from the Rhineland of Europe during colonial times. There are thus no 'old families' who occupy or claim to occupy preferred niches in the community. Moreover, there are only about 30 family names in the community and the great majority of families today have one of only about a dozen family names. During the course of the last two centuries these few families have intermarried constantly, so that the community is now one large 'Freundschaft', a term used loosely to designate kinfolk. If a member of the community is asked how many second-cousins he has, the chances are he will throw up his hands and say 'Can't count them' or 'Hundreds of them'.
>
> (Kollmorgen, 1942, p. 76)

This facet of the Old Order community has not changed appreciably over the past 50 years. The Pennsylvania Amish Directory lists 34 family names among the Amish in Lancaster County. Of these, the most popular is Stoltzfus, accounting for 25% of the total, followed by King, Fisher, Beiler and Lapp (Hostetler, 1990, p. 244).

A strong emphasis on egalitarianism also worked to prevent the formation of a community hierarchy, and consequent divisions along income and status lines. Pride, which the Amish refer to as *Hochmut*, and the striving for prestige and power to which it gives rise, is viewed as a threat to community cohesion and religious harmony. Each congregation adheres to an unwritten charter, the *Ordnung*, which regulates almost every aspect of behaviour. In addition to defining what is worldly and sinful, the *Ordnung* defines the limits of individual striving:

> ... the *Ordnung* provides a means of managing the natural human tendency toward self-exaltation (*Hochmut*) and manipulative power. Through individual submission (*Gelassenheit*) to the community's will, members are able to contribute to a network of community relationships. Tendencies viewed as disruptive and dangerous – such as self-seeking, personal power, wealth, and status – are channeled into a social order of love and brotherhood.
>
> (Hostetler, 1990, p. 84)

Of course, not all forms of inequality were eliminated. As Kollmorgen (1942, p. 76) explained:

> A member who owns a good farm is looked up to, particularly if he has a large barn and house. Some farmers are admired because they have large herds of dairy cows, others for their successful feeding operations. A farmer who maintains his farm in a high state of fertility is respected and farmers who put a great deal of manure on the land usually have prestige.
>
> The farmer 'who gets ahead' (pays off debts and buys farms) is much admired. Farmers who succeed in expanding their operations and have desirable social qualities are generally leaders in the community. It is these people whose advice is sought in agricultural matters and it is they who are generally considered for positions of leadership in the church. This is particularly true of men who, in addition to their successful farming, show certain spiritual qualities and an interest in church matters.

These distinctions, while they marked some members as more successful than others, were not pronounced enough to give rise to the kind of well-defined class structure that limits interaction between different groups. Moreover, the prestige associated with a successful farming operation was often limited to a generation or two. As family fortunes varied, so did prestige, preventing the solidification of a hereditary class structure.

Although the most visible manifestation of the Old Order Amish community's horizontal pattern is the barn-raising, it also is exemplified by the mutual aid system described above, and cooperative efforts to harvest grain, fill silos and make hay (Kollmorgen, 1942, p. 56). However, as Kollmorgen pointed out, the increased commercialization of farming activities was undermining these informal patterns of interaction. In recent years, the proliferation of ties to the market economy have further eroded subsistence-oriented cooperative activities. At the same time, however, the growth of Amish-owned businesses has created a new set of relationships. When agriculture was the dominant occupation, most retail establishments were owned by the English. Today the Amish can purchase many goods and services from fellow church members, which tends to increase day-to-day interaction within the community (Kraybill, 1989, p. 214). As an Amish grandmother put it:

> When I was little we had to go to English stores for everything, to have our implements repaired, for groceries, for dry goods, for everything. There were no Amish stores or shops then. But now we can buy everything we need in

Fig. 8.6. Old Order Amish, Pennsylvania, 1942. Sorting tobacco – a high-value cash crop.

> Amish stores – even those little calculators, we can get in an Amish store. Groceries, hardware, underwear, mittens – you name it – we can buy in an Amish store.
> (Anonymous, quoted in Kraybill, 1989, p. 214)

These stores now serve as important settings for the exchange of news and gossip and provide an opportunity for interaction between members with increasingly diverse occupations.

Stability of status and prestige markers have also worked to prevent the fragmentation that might otherwise arise within a more diverse occupational structure. As was the case when Kollmorgen visited Lancaster County, farming is still the most highly regarded activity, followed by self-employment in a non-farm business, and day labour (Hostetler, 1990, p. 140). One of the most important differentiations among Amish men continues to be good management, which includes the same qualities mentioned by Kollmorgen (1942). For the most part, however, the goals of unity and brotherly love continue to suppress the recognition of individual differences (Hostetler, 1990, p. 141).

Coincidence of Service Areas

Coincidence of service areas simply refers 'to the extent to which the service areas of local units (stores, churches, schools and so on) coincide or fail to coincide' (Warren, 1972, p. 13). At one extreme are those communities in which all service areas coincide. In such a situation, residents can be served by institutions from the same community. At the other extreme, which is

Fig. 8.7. Old Order Amish, Pennsylvania, 1995. Exterior picture of a tack and bridle shop on an Old Order Amish farm; evidence of non-farm on farm alternative income sources.

characteristic of most communities, there is much less coincidence of service areas. Residents typically travel to different locales to meet their daily needs, shopping in one community, attending church in another, and so forth.

Because the Old Order Amish travel primarily by horse and buggy, their range of movement is necessarily restricted. Thus, it would seem reasonable to assume that service areas would coincide to a considerable extent. In 1940, Leacock Township served, at least for church purposes, as the centre of the larger Lancaster County Old Order Amish community:

> Although there are no fixed focal points in the Old Order Amish Community, the whole community does have a point of concentration. When church officials from the whole community wish to meet they frequently select a home in Leacock Township which for many years has been considered the heart of the community. Intercourse is also a popular 'pairing off' place on Sunday evenings for boys and girls to go to the singings.
>
> (Kollmorgen, 1942, p. 10)

In other aspects of life, however, community boundaries were considerably more permeable. For instance, it was not uncommon for people to travel to other townships and church districts to visit with kin and friends: in social activities church districts are practically ignored. Friends and kinfolk visit back and forth regardless of location. Moreover, for the 'singings' of the young people every Sunday evening, there are no cellular divisions in the community (Kollmorgen, 1942, p. 10).

In economic matters, physical boundaries were even less important. Even though many daily needs were met by institutions within the eight-town area, the Amish had long relied on other areas of the county for goods and services and made frequent trips to Philadelphia and Lancaster City:

> Amish farmers still drive horses but they also ride in automobiles occasionally, and travel by bus and train ... [T]he Old Order Amish of Lancaster County have long been rather closely associated with cities in economic activities. The older men remember the Conestoga wagons ... [that] made numerous trips to haul surplus products to market in Philadelphia and to bring back home and neighborhood needs or cargo for merchants in or near the City of Lancaster. [A]fter the years of the Conestoga wagons, many ... people made weekly trips to the market in Lancaster or some other place. Some ... still do. Moreover, sickness and shopping bring these people to Lancaster at intervals.
> (Kollmorgen, 1942, p. 92)

Development pressure and a burgeoning population have simultaneously concentrated and dispersed service areas. On the one hand, the scarcity of available farmland and high land prices have forced many Amish who want to remain in farming to settle in other parts of the county. Many of these areas are too far away from the centre of the settlement to shop at Amish stores, or for friends and relatives to visit one another easily by horse and buggy. Consequently, visits are more infrequent and must be made by car or van. On the other hand, the number of Amish-owned retail and service businesses has increased dramatically in Kollmorgen's original study area. Thus, for the Amish living in this locality, many goods and services previously available only in Lancaster or other communities can now be bought locally.

Fig. 8.8. Old Order Amish, Pennsylvania, 1995. Typical means of transportation remains a buggy, all of which carry a 'slow-moving traffic' warning sign. Despite these signs and the common presence of buggies on highways, accidents between other vehicles and buggies are common.

Psychological Identification

It is difficult to disentangle psychological identification with a locality from the other dimensions of community life discussed throughout this chapter. Autonomy, horizontal ties and coincidence of service areas all contribute to a sense of belongingness. Hence, it is probably more reasonable to view psychological identification as an outgrowth of the first three dimensions. Moreover, community variations in autonomy, horizontal pattern and coincidence of service areas are likely to be associated with variations in the extent to which residents identify with a common locality. In places with a great deal of autonomy, a strong pattern of horizontal ties and coincidental service areas, there is likely to be a strong sense of community identity. Conversely, places that lack autonomy, display a weak pattern of horizontal ties and are characterized by widely dispersed service areas are unlikely to engender strong feelings of commonality among residents. Thus, psychological identification can, like the other dimensions of community life, be depicted as existing along a continuum ranging from weak to strong.

Discussions of psychological identification among the Old Order Amish are further complicated because they require a distinction between identification as a unique religious group and identification with the locality. There is little doubt that the Old Order Amish in the Lancaster Study area were, in 1940, a distinct, homogeneous group. Their religion, their language, their clothing, their internalized values and their almost universal occupation of farming contributed to a highly developed sense of group identity.

Kollmorgen (1942, p. 4) began his description of the Amish by stating: 'The garb and mode of life of the Amish differentiate them sharply from other people. The Amish consider themselves a 'peculiar' people who lead a 'peculiar' life because the Bible says that God's people are peculiar and are not conformed to the world.' The men's beards and their long, banged, parted-in-the middle hair immediately marked them as Amishmen. Each age and sex group had their own distinctive dress and hair characteristics. These features identified them as Amish to the 'English'. More importantly, they served as constant reminders to themselves of their religious history and the principles of nonconformity and the unequal yoke. The dress code also prevented them from buying 'ready-made dresses, overalls, shirts, men's suits, or even dress overcoats' (Kollmorgen, 1942, p. 47).

Describing the extent to which the Old Order Amish identified with the study area or Lancaster County was more difficult. In fact, Kollmorgen (1942) did not address this issue directly. Although there are several references to the 'Amishman's attachment to his land' (Kollmorgen, 1942, p. 31), no attempt was made to generalize about feelings of attachment to, or identification with, the larger locality. However, his discussion of folklore and its role in transmitting wisdom and knowledge through the generations highlights some key features of a unique heritage narrative (Bridger, 1996) which linked the Amish to Lancaster County through an account of the community's origins, the character of its people and its trials and triumphs over time:

> The aged farmers in the Amish and Mennonite communities in Lancaster County are a rich depository of the lore of agricultural experience which reaches back to early colonial days. Many of them, particularly in the Mennonite community, occupy farms that have been in the family for more than 200 years and were acquired directly from William Penn or his land agents. Buildings have been built, at least partly, shortly before or after the Revolutionary War. As the farm place was handed down from father to son, so were many experiences handed down from generation to generation and with them a good deal of wisdom. Many an old-timer remembers the recurrent depressions in the latter half of the 19th century which were occasioned by the financial instability and the severe competition with the emerging granary of the Middle West … [T]he family farm place and many of the family heirlooms focus attention on the activities and experiences of the forefathers.
>
> (Kollmorgen, 1942, p. 26)

A more complete version of this narrative would include an account of Anabaptist origins during the Reformation, persecution and suffering in Europe, the voyage to America, the search for a settlement site, the important role that land stewardship, careful management and thrift have played in developing a successful community, the persecution and suffering that has helped bind the community, and so forth. As this story was handed down from generation to generation, it not only strengthened identification with the group but also taught children about their longstanding and important relationship with Lancaster County.

To a remarkable degree, the religious cohesion that characterized the Old Order Amish of Lancaster County in 1940 has persisted. Unlike many Amish settlements in other parts of the country, the Lancaster County Old Order Amish have experienced relatively little fragmentation.[5] The most serious split occurred in 1966 when a group called the 'New Amish' began to emerge (Hostetler, 1980, p. 277). This movement was driven by families who favoured the use of telephones, tractor-driven farm machinery and power-driven generators for cooling bulk milk. Adherents also spoke out against the use of tobacco and called for 'a cleaner lifestyle and conduct' (Hostetler, 1980, p. 277). Like their Old Order counterparts, the New Amish continue to meet for worship in one another's homes.

Many Amish, however, see current trends as a threat to the principle of unequal yoke that stands at the core of their religious belief. The growing number of Amish working off-farm is a particularly troubling issue to church leaders. As Testa (1992, p. 61) puts it,

> To be a farmer who earns his primary livelihood on a construction crew building tract houses, to be a waitress in restaurant or a clerk in a store, to hang a sign outside one's farm … advertising the sale of quilts or wooden lawn furniture is to be 'yoked' to the outside world and fundamentally at odds with the Amish faith.

There also are fears that off-farm employment will erode the homogeneity of experience that has been central to the Amish way of life. When virtually all community members were engaged in agriculture and closely related occupations, it was much easier to maintain a common identity:

> It is extremely difficult for a group to remain 'peculiar' in the crowded life of cities. The influences of the urban centers on their chosen way of life are well realized by the Old Order Amish, and so the occupation of farming or some closely associated activity in rural or semi-rural areas is now a test of church membership.
>
> (Kollmorgen, 1942, p. 23)

Agriculture enhanced both physical and social separation from the larger society. The growth of off-farm employment, and the unprecedented diversity of life experience it generated, have led some Amish to question the long-term implications for group identity. An Amish elder summarized this view of the future well when he said, 'Soon there'll be two Amish groups here. There'll be farmers and the trade people. And that will be the end of us' (Anonymous quoted in Testa, 1992, p. 61).

Amish ties to Lancaster County (or any other locality for that matter) have always been secondary to their desire to preserve their religious and cultural integrity. In fact, before coming to America, periodic rounds of persecution forced the Amish to roam extensively over Alsace, Lorraine, Palatinate, Baden and Switzerland. After moving to Lancaster County and settling into a predictable, agrarian lifestyle, psychological identification with the 'Garden Spot' was largely taken for granted:

> The Amishman's attachment to his land is perhaps exceeded only by his attachment to his religion. It follows that there is little or no inducement to relinquish his land as long as his socio-religious life is agreeable ... When intolerable infringements are imposed from without, however, the Amish, like the Mennonites generally, are prepared to make great sacrifices. Such infringements have resulted in numerous major migrations of nonresistant people during the last few centuries.
>
> (Kollmorgen, 1942, p. 31)

To many Amish, the current difficulty of securing a livelihood from farming constitutes the kind of infringement that would warrant migration. This view was already gaining currency in the 1970s, when an Amishman interviewed by the *New York Times* discussed the problems associated with finding farms for his children:

> That's our biggest problem, to keep our children on the farm. There's just no more room to grow. Lancaster County is my home. All the roots are here. Leaving would be like pulling up a plant by the roots. But I would do that, if that's what it takes, before we give up our faith. I would rather my children leave the county than see them work in a factory.
>
> (Anonymous, quoted in Ericksen *et al.*, 1980, p. 660)

Talk of migration increased in the 1980s and 1990s, as land prices soared and non-Amish farmers who wished to leave farming found that they could command a higher price from developers than from the Amish. Scouting parties have been dispatched to search for suitable farmland in Kentucky, the Finger Lakes region of New York state, and as far west as Indiana and Wisconsin (Testa, 1992, p. 144). To date, most Lancaster Amish have remained in the county, and what migration there has been has largely been confined to areas within Pennsylvania (Hostetler, 1990, p. 371). However, if the county's current growth rate continues, greater out-migration is possible.

Conclusion

To the millions of tourists who visit Lancaster County each year, the Old Order Amish seem to be a society frozen in time. In its promotional materials, the Pennsylvania Dutch Visitors Bureau depicts the county as a place where the best of 19th-century rural America lives on – magazine and television advertisements feature pictures of the Amish clip-clopping down country lanes in their horse-drawn buggies, working neatly plowed fields with old-fashioned implements, and joining with their neighbours in communal barn raisings.

While there is an element of truth to each of these scenarios, they mask the rather significant transition made by the Old Order Amish over

Fig. 8.9. Old Order Amish, Pennsylvania, 1942. Grossdaadi Haus – Amish added on to their homes as the family got larger.

the past 50 years. With growth and development pushing in from all sides, the Amish have consistently moved away from their agrarian roots. Kollmorgen (1942, p. 104) recognized this potential and warned of the consequences for the Old Order community:

> Recognition of the fact that a rural life is essential for the survival of the church does not solve the problems that result from the regulation limiting members to rural occupations. To remain on the land bespeaks an expansion of land holdings, and such expansion brings problems in its wake. The pressure on the land in Lancaster County and resultant high prices for land have led to the establishment of new communities from time to time. It is possible that available land for new communities may be increasingly difficult to find. If and when appropriate land for community expansion cannot be found, the resistance to urban opportunities may weaken. Once factory and urban jobs are accepted the peculiar and rural ways of life are seriously threatened.

Although some of these predictions have come true and many Amish worry about the future of their community, Amish society has not disintegrated. Somehow they have managed to accommodate changes that do not threaten religious harmony or community cohesion and resist those that do. Kraybill (1989) uses the metaphor of cultural fences to describe how the Amish regulate change. As he puts it, 'Coping with social change involves fortifying old fences in some areas, as well as moving fences in other ones' (Kraybill, 1989, p. 236). Thus, church leaders might reinforce the proscription against the use of video cameras while allowing the widespread adoption of pocket calculators (Kraybill, 1989). One change (the use of video cameras) would violate the prohibition against graven images and indicate personal pride and vanity, while the other (pocket calculators) would not cut to the core of fundamental religious values.

Of course, this process is not uniform across church districts. The Amish are now too numerous and diverse to maintain universal standards. Nevertheless, there is a common element to the way in which social, economic and technological change is deliberated. Whenever there is a question about whether to allow some new innovation or departure from tradition, the matter is considered in terms of its implications for community and religious stability over the long term. An Amish craftsman, discussing the taboo on the telephone, provides a good example of this: 'If we allow the telephone, that would be just a start. People would say, "Okay, now we'll push for that ..." It would be a move forward that might get the wheel rolling a little faster that we can control it, if you know what I mean' (Anonymous, quoted in Kraybill, 1989, p. 240).

Another means of regulating change, particularly externally imposed change, has not received as much attention. With the growth of tourism, the Amish have assumed an economic importance beyond what their population would suggest. Without the Amish, tourist revenues would be only a fraction of the US$450 million generated annually. The Amish realize this and are increasingly willing to use the importance of their presence as a bargaining chip with state and local officials. As we noted above, the Amish have periodically discussed leaving Lancaster County for more rural areas. Although the possibility of large-scale out-migration cannot be dismissed out of hand, talk of moving also can be viewed as a threat which says, in effect, 'We know that the fortunes of Lancaster County are intimately bound up with the fortunes of the Amish. Tourists come to see us, and if the industry is to thrive, we must be allowed to retain those features of our lifestyle that distinguish us as a peculiar people.' Of course, the Amish would not make such a bold statement openly. Instead, it runs as an undercurrent through negotiations with the 'English', letting them know that although the Amish are a non-resistant people, they will be pushed only so far. They will not actively resist. They will simply move away and take a multi-million dollar tourist industry with them.

In Lancaster County, natural population increase in the Old Order community, the expansion of local, state, and federal regulations, and rapid growth and development have forced the Amish to confront situations that were unthinkable 50 years ago. Adjustment has involved changes within the community (especially in the division of labour) as well as a series of conflicts, negotiations and compromises between the Amish and the many 'English' institutions with which they must now interact. Whether the issue was social security, education, farming practices or land use and zoning, the Amish have, for the most part, managed to create outcomes that maintain continuity with the past. Indeed, it is their shared understanding of this

Fig. 8.10. Old Order Amish, Pennsylvania, 1995. Typical Old Order Amish Farm in Lancaster County.

past that has enabled to the Amish to confront new situations with a vision of the future that is lacking in most communities. Without this sense of group identity based on a deeply embedded religious world view, it is doubtful that the Amish would have been able to maintain a relatively autonomous, geographically compact and tightly knit community in the middle of the Washington–Boston megalopolis.

Notes

[1] Support for this research was provided by the Pennsylvania Department of Agriculture (ME 442152) and the Pennsylvania Agricultural Experiment Station (Regional Project NE 173 and State Station Project 3548). We gratefully acknowledge the efforts of the NE 173 Regional Research Technical Committee: Kenneth P. Wilkinson, Steve Jacob, Kathleen Miller and Linda Kline.

[2] For a discussion of procedures used to conduct the analyses of land-use change see Luloff and Befort (1989).

[3] Amish farms in Lancaster County have become progressively smaller over the past century. In 1900, the average size of an Amish farm was 100 acres. By 1970, many farms were 40 acres or less (Fisher, 1978, p. 111) and, because farmland in Lancaster County is now so limited and expensive, profitability hinges on intensive land use. It is not uncommon for interest payments on new farms to range between US$20,000 and US$40,000 per year (Hostetler, 1990, p. 131). To generate this much cash and turn a profit, many farmers have installed large dairy, poultry and hog operations which generate much more manure than necessary for fertilization.

[4] The Fish and Boat Code contains a littering provision that makes it unlawful to discharge any substance 'in such a manner that it is carried into the water' (Place, 1993, p. 195). According to Place, this statute is the basis for most enforcement actions against farmers in Pennsylvania.

[5] A stark contrast to the relative homogeneity of the Old Order Amish in Lancaster can be found in the Amish settlement in Kishacoquillas Valley in central Pennsylvania. Orginally settled by Amish who came from south-eastern Pennsylvania in 1791, Kishacoquillas Valley eventually split into 12 distinct Amish-related groups. The groups differ primarily by the extent to which they have assimilated into the larger culture, using the terms 'low' and 'high' to designate how each relates to the old traditions. 'A low church is one that has retained the old traditions, while a high church is one that is more like the world' (Hostetler, 1980, p. 282).

Appendix

Table A1. Sociodemographic characteristics for eight Lancaster County townships, 1940.

	Upper Leacock	Leacock	West End	Earl	East Earl	Salisbury	East Lampeter	Paradise
Population	2964	2329	2663	2875	2638	3579	3847	2574
Square miles	18.1	20.6	17.6	21.9	24.6	41.9	19.9	18.7
Persons per square mile	164	113	151	131	107	85	193	138
Per cent rural-farm	43.9	72.5	54.4	59.8	45.3	51.6	36.8	35.3
Per cent non-white	0.00	0.00	0.00	0.00	0.01	0.03	0.01	0.00
Per cent over 64	7.6	7.1	5.4	5.4	7.8	13.3	8.0	8.5
Per cent less than 14	30.4	36.5	34.5	37.0	35.2	30.7	29.8	28.4
Per cent female	51.7	52.7	49.8	49.2	49.4	47.9	51.9	50.1
Per cent over 25 completed high school[a]	18.4	13.2	7.9	8.6	14.5	12.6	23.8	13.2
Per cent over 25 with bachelor's degree[a]	4.1	3.3	1.8	1.5	2.5	2.1	5.8	3.3
Median family income[a] (US$)	5364	4594	5050	3842	5379	4355	5787	4593
Per cent families in poverty	5.9	6.6	7.4	12.4	4.0	9.4	3.9	6.6
Per cent in labour force[a]	57.6	54.7	61.0	57.6	58.4	57.5	60.1	54.7
Per cent female in labour force[a]	34.7	28.7	38.0	27.8	36.0	32.1	36.6	28.7
Per cent employed in manufacturing[a]	31.3	29.6	43.8	27.6	42.2	27.9	34.1	29.6
Per cent employed in service[a]	4.0	3.8	5.2	1.3	2.5	4.7	7.9	3.8
Per cent owner-occupied housing	54.7	57.2	55.6	58.4	69.9	64.1	57.8	58.3
Per cent renter-occupied housing	44.6	42.8	37.5	37.5	29.3	33.3	42.3	41.8

[a]Due to lack of 1940 data at the minor civil division level, 1960 census tract data was used instead.

Table A2. Sociodemographic characteristics for eight Lancaster County townships, 1990.

	Upper Leacock	Leacock	West End	Earl	East Earl	Salisbury	East Lampeter	Paradise
Population	7254	4668	6434	5515	5491	8527	11,999	4430
Square miles	18.1	20.6	17.6	21.9	24.6	41.9	19.9	18.7
Persons per square mile	417	236	364	244	244	217	606	246
Per cent rural-farm	11.2	18.0	10.4	18.9	18.4	12.4	4.0	14.7
Per cent non-white	3.6	1.0	2.3	1.3	1.1	1.8	3.6	0.0
Per cent over 64	11.3	11.0	11.8	10.6	10.2	8.9	13.8	10.2
Per cent less than 14	25.1	32.6	26.5	29.3	28.8	28.6	19.9	27.2
Per cent female	49.5	49.9	51.0	51.1	50.4	49.5	50.8	49.7
Per cent over 25 completed high school	61.4	33.6	55.6	46.5	48.7	50.9	72.5	59.7
Per cent over 25 with bachelor's degree	10.6	6.9	14.0	10.6	4.0	8.0	19.0	9.2
Median family income (US$)	31,436	29,525	36,121	33,003	32,030	31,893	34,720	34,762
Per cent families in poverty	8.1	13.0	5.7	6.6	6.4	7.8	3.3	8.6
Per cent in labour force	72.3	68.1	72.9	68.0	69.3	70.4	71.6	70.3
Per cent female in labour force	59.3	51.2	60.2	51.1	52.8	57.4	61.5	59.5
Per cent employed in manufacturing	15.3	18.5	13.0	16.8	19.1	22.8	13.1	17.6
Per cent employed in service	13.4	15.0	9.7	10.4	10.8	14.6	12.7	11.1
Per cent owner-occupied housing	59.8	66.8	80.7	75.0	75.3	75.7	64.0	25.6
Per cent renter-occupied housing	37.2	30.8	16.8	22.4	22.0	21.7	32.4	25.6

A 50-year Perspective on Persistence and Change: Lessons from the Rural Studies Communities

Richard S. Krannich and A.E. Luloff

When the six communities examined in the preceding chapters were first studied in the early 1940s, rural America was in the throes of an extraordinary period of change and upheaval. During the first three decades of the 20th century, the effects of urban industrial expansion and the initial impacts of the mechanization of agriculture combined to spur a major shift in migration and settlement patterns. As a result, large numbers of rural residents moved to urban centres in search of improved economic opportunity and a better quality of life. The Great Depression of the 1930s hastened the pace of economic and demographic destabilization in many parts of rural America. By the time the original series of Rural Studies were begun, a pattern of rural community decline and instability was broadly evident across much of the American landscape.

Although the original selection of the Rural Studies communities was based on their positions along an instability–stability continuum, evidence of substantial change, and in some cases severe instability, was evident across these study sites by the time the initial research was completed. Research design and study selection decisions made by the original investigators, which may have deviated somewhat from the formal guidelines established at the initiation of the Rural Studies project, could account for the evidence of flux and instability across virtually all of the study areas. However, it is also likely that this reflects the dynamics of community change occurring throughout rural America at the time. In that context, even places that may have appeared to be relatively stable based on preliminary reconnaissance when the studies were initiated, were likely to prove far less stable upon more in-depth investigation over several ensuing years.

Among the study sites, major economic, technological and structural transformations in agriculture were associated with substantial depopulation in Irwin, Landaff, Harmony and Sublette. Even among the two cases conceived as being the most stable, the Old Order Amish of Lancaster County (who were characterized by their highly conservative and sectarian religious culture) and the residents of El Cerrito (who clung to a set of deeply entrenched cultural and religious traditions), evidence presented in the original Rural Studies reports suggested that the broader forces of economic and social change were threatening to engulf and transform them as well. As with numerous studies of American rural communities conducted at mid-century, the findings reported in the original Rural Studies reports provided little to suggest that an 'eclipse' of local communities was not, in fact, well under way and, generally, inescapable.

The pace and magnitude of economic, demographic and social changes that enveloped these communities during the decades following the original studies certainly did not abate. In areas of Lancaster County encompassing the Old Order Amish community, the forces of population growth and growing competition for non-agricultural land uses have transformed the surrounding rural landscape and forced some members of the Amish community to relocate their farms and families to distant areas. Increased linkages to the economic and political institutions of the broader society have generated heightened concerns about the autonomy, stability and sustainability of the Amish community and its unique religious and cultural traditions.

El Cerrito, also considered stable in 1940, subsequently experienced a nearly complete loss of population as families abandoned their homes and moved to distant urban centres in search of employment opportunities. Harmony, which in 1940 had endured the effects of racial divisions and the loss of nearly half of its population (who were largely African-American) due to the collapse of cotton production and the plantation system, was partially obscured during the 1970s by the development of a large dam and reservoir. Landaff witnessed continued population declines for 40 years following the initial study period, along with a nearly complete disappearance of the dairy farms that had once sustained the local population. In Irwin and the surrounding Shelby County area, agricultural transformations over the past 50 years contributed to dramatic decreases in the number of farms, substantial depopulation of open-country areas, an increasingly aged in-town population and a withering of many local businesses. In contrast, Sublette and surrounding portions of Haskell County, considered to be the least stable of the study settings in 1940, experienced a different pattern of transformation as evidenced by a shift from dramatic population loss to a pattern of sustained population growth associated with an increasingly industrialized and vertically integrated system of irrigated agriculture and expansion of energy resource industries in the area.

The key finding that emerges from our restudy of these six communities, a half-century after the initial effort, revolves around the somewhat amazing persistence of localized social organization, community attachment and community action – regardless of the disparate and transformed settings characteristic of these places. In no case is there evidence that the social fabric of community collapsed in the face of population change, economic transitions or increased linkages to extra-local organizations and processes. Certainly, each community is dramatically different from what it was in 1940. All have witnessed the effects of declining local autonomy and increased vertical linkages to external organizations and authorities that accompanied the 'Great Change' process outlined by Warren (1978). The patterns of change evident in these places, while of paramount importance in understanding the great diversity of conditions and adaptations that confronted residents of rural America, should not, however, be presumed to reflect the so-called 'eclipse' of community that many earlier community analysts had anticipated.

Indeed, our restudies revealed a degree of persistence, and even vibrancy, indicating that community was alive and well in varying ways across all six study areas. This is evidenced by major community events and celebrations, such as Old Home Days in Landaff, the annual *limpia* event that revolves around the clearing and repairing of the irrigation ditch in El Cerrito, and the Centennial Celebration in Irwin.

It is also evident in a broad range of collective action episodes indicative of localized efforts to address shared needs and concerns. Examples of such collective action include the community response in Irwin to assist a local family who lost their café and attached dwelling in a fire, the traditional barn-raising activities of the Old Order Amish, the collective efforts to repair a flood-damaged dam and diversion ditch in El Cerrito, the emergence of a 'Concerned Citizens Group' to address perceived tax assessment inequities in Putnam County, and broad-based community involvement in efforts to sustain the 'Blue School' in Landaff. Substantial evidence was presented to suggest that most local residents evinced high levels of community attachment and identification. Further, in each study site, there was a substantial commitment to sustaining local institutions and cultural traditions.

It is also obvious that the factors that contributed to the persistence of community in the face of major economic, demographic and social upheavals were multifaceted and highly variable across the study sites. First, the persistence of community was clearly intertwined with the cultural traditions and belief systems that, to varying degrees, characterized these settings. While such forces were most evident in the culturally unique contexts of the Old Order Amish and El Cerrito, evidence of the importance of local culture and heritage could be drawn from each of the other study settings as well.

Also evident were the increasingly important roles that a variety of public and non-public institutions played in the persistence of community across these contexts. For example, the role of localized government was extremely important in Landaff, where a strong adherence to the tradition of community-based governance promoted broad-based citizen involvement in local affairs. Local school systems represented another public institution that frequently played a central role in encouraging community involvement and action, as was particularly evident in the case-study findings for Irwin and Landaff. In some settings, the community fire department played a similar role, as in Landaff. In El Cerrito, where public institutions were otherwise virtually absent, the *Acequia* Association played a crucial role in fostering local action and coordinating ties with external authorities. Elsewhere, religious culture and local churches played a central role in maintaining community norms and coordinating community involvement, as reported in the chapters on Sublette, Irwin and Harmony.

Also important in accounting for community persistence were the effects of a broad range of governmental programmes and policies that helped to stimulate the development of new economic opportunities or to fund community facilities and services necessary to meet the needs of local populations. In Sublette, federal farm programmes played a central role for many years in the stabilization and expansion of the local and regional economy. Government funding of the Lake Oconee dam project in Putnam County simultaneously obliterated a portion of the traditional African-American Harmony community and stimulated a substantial in-migration of primarily White and economically well-off new residents and property owners and associated economic expansions that have been central to the surrounding area's recent development patterns. Federal and state funding of a culinary water system, bridge replacement and diversion dam reconstruction in El Cerrito were crucial to the survival and gradual re-emergence of that community.

Finally, the spatial relationships of rural communities with nearby or regional urban population centres, and the effects of such locational features on the attractiveness of these places as recreation sites, commuter residence locations, or destinations for persons migrating toward more rural settings, have had important impacts on community change and persistence. For example, the substantial demographic transformations evident in the Harmony/Putnam County study area were closely linked to its proximity to the Atlanta metropolitan area, and its growing popularity as a location for both seasonal and recreational homes and for ex-urban residential developments. Despite a degree of spatial isolation, El Cerrito's survival was clearly linked to the ability of its residents to access both services and economic opportunities in larger urban places in northern New Mexico, including Las Vegas and Santa Fe. Many of the economic ventures and employment opportunities that helped to sustain members of the Old Order Amish community were a consequence of that community's location in the rapidly growing and increasingly urbanized setting of Lancaster County. In contrast, Landaff's relative isolation from major urban centres helped to stimulate the in-migration of new residents who were purposefully seeking a quality of life tied to a relatively traditional rural community setting. In cases where community change involved the in-migration of new residents, there was evidence of increased economic activity, an infusion of new human capital into the local arena, and, in some instances, a stimulation of social interactions and organizations needed to more successfully pursue community actions. The arrival of new populations also created tensions and divisions, as evident from both the Harmony/Putnam County and Landaff restudies. However, in the absence of such population growth, there is little doubt that those places would be both less dynamic and less capable of responding to local interests and needs than our restudies indicated.

Our restudies demonstrate that the six study communities have persisted, despite the dramatic changes that occurred during the latter half of the 20th century. While continued changes in economic, demographic and social conditions are virtually certain to occur across each of these settings, and in rural America more broadly, the resilience of sentiments, interactions and community activeness evident from our analyses suggests that successful adaptation is likely to occur. It is important, however, to acknowledge the difficulties and challenges that confront many American rural communities and that continue to threaten their stability and sustainability. As Wilkinson observed, the emergence of localized interactions and collective action episodes central to the concept of community does not reflect a contrived process, but instead reflects a 'natural disposition among people' to 'engage in social relationships with others' and 'have their social being and identity in that interaction' (Wilkinson, 1991, p. 111). At the same time, it is not easy to sustain community, because the conditions of modern rural America often generate barriers to the natural emergence of such associations and actions. To varying degrees, each of the communities examined here confront challenges associated with the absence of a comprehensive array of employment opportunities, social organizations and services needed to fully meet the needs and interactional interests of local residents. When economic opportunities and necessary facilities and services are not available in the locality, people must either relocate to areas where those needs can be met or spend much of their time and energy in interactions outside of the locality as they pursue those needs in other places. Under such circumstances, the potential for localized interaction and the capacity for collective action is inevitably diminished.

Moreover, each of the communities examined here exhibits conditions that could contribute to future destabilization and a further erosion of community capacity to meet local needs and ability to collectively respond to shared interests and concerns. The potential for major economic decline was evident in places like Sublette, which exhibited a high degree of dependence on a highly volatile energy industry and on agricultural irrigation practices that were not sustainable over the long term. In Harmony, dependence on low-wage industries, such as textile manufacturing and services associated with recreational development, created different forms of vulnerability, particularly since such industries can become highly volatile in the face of global restructuring of industrial location patterns as well as national and global economic fluctuations. Continued land development pressures, coupled to increased government restrictions on farming practices affecting water quality or other environmental concerns, could, in the future, force a substantial displacement of the Old Order Amish community. Government-imposed restrictions on irrigation water withdrawals could severely undermine social and economic conditions in both El Cerrito and Sublette. Continued erosion of locality-based employment, services and organizations has the potential to constrain local involvement, attachment and action potential in each of the six settings, but seems especially challenging for places like Irwin and Landaff, where the pace of consolidation and concentration of infrastructure and services in other regional communities is of particular concern.

In some ways, our restudy found these sites to be confronting the same difficulties that the authors of the original studies noted. We also experienced some of the same research challenges. Like them, we found that the pursuit of a single theoretical or methodological template to standardize analysis across each distinct setting was impossible to accomplish, both because of the differing interests and inclinations of individual investigators and because the unique contexts of the communities mandated differing approaches and emphases. However, unlike the initial group of researchers, we benefited greatly from the availability of in-depth reports on community conditions at a prior point in time, and the ability to draw upon those analyses to examine and assess long-term patterns of change.

Indeed, we believe that a major contribution of our restudy effort is the fact that it highlights the need for both sociologists and policy analysts to pursue longer-term perspectives and visions regarding the conditions affecting rural people and communities. Our analyses, and the comparisons of contemporary conditions with those evident at the time of the initial Rural Studies investigations, indicated that a snap-shot glimpse of any of these communities at a single point in time would potentially be highly misleading with respect to patterns and trends in stability–instability (as well as most other conditions of local life). The community considered least stable in 1940 (Sublette) actually exhibited more economic and demographic stability, and even growth, in the ensuing years than any of the others examined as part of this effort. Conversely, one of the 'most stable' communities in 1940 (El Cerrito) exhibited a virtual collapse in subsequent years, only to experience an unexpected rebound. Similarly, relatively contemporary studies of rapid community change associated with the volatility of natural resource-based industries generally offer little in the way of conclusive evidence regarding the effects of such industries and changes on social well-being and community capacity in a number of rural locales. In major part, this failure reflects the absence of longitudinal study designs in most studies of such settings (however, see Freudenburg and Gramling, 1992; Smith *et al.*, 2001). A full understanding of contemporary community conditions can only be accomplished through consideration of past conditions and the patterns of change and adaptation that occur over extended time periods. Similarly, efforts to anticipate what is 'coming down the pike' with respect to future community change must be informed by an understanding of how present conditions are a reflection of past change.

References

Aiken, C.S. (1990) A new type of black ghetto in the Plantation South. *Annals of the American Geographers* 80, 223–246.

Allen, J.C. and Dillman, D.A. (1994) *Against All Odds: Rural Community in the Information Age.* Westview Press, Boulder, Colorado.

Anderson, S. (1960) *Winesburg, Ohio.* Viking Press, New York.

Bates, F.L. and Bacon, L. (1972) The community as a social system. *Social Forces* 50(3), 371–379.

Bauer, R.A. and Bauer, A.H. (1960) America, 'mass society' and mass media. *Journal of Social Issues* 16(3), 3–66.

Beggs, J.J., Hurlbert, J. and Haines, V. (1996) Community attachment in a rural setting: a refinement and empirical test of the systemic model. *Rural Sociology* 61(3), 407–426.

Bell, D. (1956) The theory of mass society. *Commentary* July, 75–83.

Bell, E.H. (1942) *Culture of a Contemporary Rural Community: Sublette, Kansas.* Rural Life Studies 2. USDA, Bureau of Agricultural Economics, Washington, DC.

Bridger, J.C. (1996) Community imagery and the built environment. *The Sociological Quarterly* 37(3), 353–374.

Bridger, J.C. and Luloff, A.E. (1999) Toward an interactional approach to sustainable community development. *Journal of Rural Studies* 15(4), 377–387.

Bridger, J.C. and Luloff, A.E. (2001) Building the sustainable community: is social capital the answer? *Sociological Inquiry* 71(4), 458–472.

Brunner, E. and Kolb, J.H. (1933) *Rural Social Trends.* McGraw-Hill, New York.

Brunner, E. and Lorge, I. (1937) *Rural Trends in Depression Years.* Columbia University Press, New York.

Brunner, E., Hughes, G.S. and Patten, M. (1927) *American Agricultural Villages.* George H. Doran, New York.

Buck, R.C. (1978) Boundary maintenance revisited: Tourist experience in an Old Order Amish community. *Rural Sociology* 43(2), 221–234.

Caplow, T.H., Bahr, H.M., Chadwick, B.A., Hill, R. and Williamson, M.H. (1982) *Middletown's Families: Fifty Years of Change and Continuity.* University of Minnesota Press, Minneapolis, Minnesota.

Caplow, T.H., Bahr, H.M., Chadwick, B.A., Hoover, D.W., Martin, L.A., Tamney, J.B. and Williamson, M.H. (1983) *All Faithful People: Change and Continuity in Middletown's Religion.* University of Minnesota Press, Minneapolis, Minnesota.

Carter, R.M. (1946) *The Development and Financing of Local Governmental Institutions in Nine Vermont Towns.* Bulletin 529, AES, University of Vermont, Burlington, Vermont.

Carter, R.M. (1947a) *The People and Their Use of Land in Nine Vermont Towns.* Bulletin 536, AES, University of Vermont, Burlington, Vermont.

Carter, R.M. (1947b) *Rural Non-Farm Family Living in Nine Vermont Towns.* Bulletin 537, AES, University of Vermont, Burlington, Vermont.

Castells, M. (1983) *The City and the Grassroots: a Cross-Cultural Theory of Urban Social Movements.* University of California Press, Berkeley, California.

Conner, D. and Schmidt, F. (1996) Analysis of Community Attachment: Landaff, New Hampshire. Study of Community Change and Persistence, Regional Project NE-173. Paper presented at the Rural Sociological Meetings, Iowa, 1997. Center for Rural Studies, University of Vermont, Burlington, Vermont.

Cox, K.R. and Mair, A. (1988) Locality and community in the politics of local economic development. *Annals of the Association of American Geographers* 78, 307–325.

Crawford, S. (1988) *Mayordomo: Chronicle of an Acequia in Northern New Mexico.* University of New Mexico Press, Albuquerque, New Mexico.

Cronon, W. (1991) *Nature's Metropolis: Chicago and the Great West.* W.W. Norton, New York.

Currier, S.P. and Clement, E.T. (1966) *History of Landaff New Hampshire.* Courier Printing, Littleton, New Hampshire.

Davis, A., Gardner, B.B. and Gardner, M.R. (1944) *Deep South.* University of Chicago Press, Chicago, Illinois.

Denzin, N.K. (1989) *The Research Act: a Theoretical Introduction to Sociological Methods*, 3rd edn. Prentice-Hall, Englewood Cliffs, New Jersey.

Des Moines Register (2000) Fading Rural Areas Feed Cities. 13 March.

Des Moines Register (2000) Rural, Metro Iowa at Odds. 16 March.

Dewey, R. (1960) The rural–urban continuum: real but relatively unimportant. *American Journal of Sociology* 66(1), 60–66.

Durkheim, E. (1947) *The Division of Labor in Society.* (Translated by George Simpson.) The Free Press, Glencoe, Illinois.

DuWors, R. (1952) Persistence and change in values of two New England communities. *Rural Sociology* 17(3), 207–217.

Eastman, C. and Krannich, R.S. (1995) Community change and persistence: the case of El Cerrito, New Mexico. *Journal of the Community Development Society* 26(1), 41–51.

Eastman, C. and Krannich, R.S. (1999) *El Cerrito: a photo essay.* New Mexico State University,

Agricultural Experiment Station, Las Cruces, New Mexico.

Edwards, A.D. (1939) *Influence of Drought and Depression on a Rural Community: a Case Study of Haskell County, Kansas.* Social Research Report No. VII, United States Farm Security Administration, Washington, DC.

Ericksen, E.P., Ericksen, J.A. and Hostetler, J.A. (1980) The cultivation of the soil as a moral directive: population growth, family ties, and the maintenance of community among the Old Order Amish. *Rural Sociology* 45(1), 49–68.

Eriksen, J. and Klein, G. (1981) Women's roles and family production among the Old Order Amish. *Rural Sociology* 46(2), 282–296.

Ferrera, P.J. (1993) Education and schooling. In: Kraybill, D.B. (ed.) *The Amish and the State.* The Johns Hopkins University Press, Baltimore, Maryland, pp. 87–108.

Fisher, G.L. (1978) *Farm Life and Its Changes.* Pequea Publishers, Gordonville, Pennsylvania.

Freudenburg, W.R. (1986) The density of acquaintanceship: an overlooked variable in community research? *American Journal of Sociology* 92(July), 27–63.

Freudenberg, W.R. and Gramling, R. (1992) Community impacts of technological change: toward a longitudinal perspective. *Social Forces* 70, 937–955.

Gallaher, A., Jr (1961) *Plainville Fifteen Years Later.* Columbia University Press, New York.

Gallaher, A., Jr (1980) Dependence on external authority and the decline of community. In: Gallaher, A., Jr and Padfield, H. (eds) *The Dying Community.* University of New Mexico Press, Albuquerque, New Mexico, pp. 85–108.

Geertz, C. *The Interpretation of Cultures.* Basic Books, New York.

Goudy, W.J. (1977) Evaluation of local attributes and community satisfaction in small towns. *Rural Sociology* 42(3), 371–382.

Goudy, W.J. (1983) Desired and actual communities: perceptions of 27 Iowa towns. *Journal of the Community Development Society* 14(1), 39–49.

Goudy, W.J. (1990) Community attachment in a rural region. *Rural Sociology* 55(2), 178–198.

Green, G.P., Marcouiller, D., Deller, S., Erkkila, D. and Sumathi, N.R. (1996) Local dependency, land use attitudes, and economic development: comparisons between seasonal and permanent residents. *Rural Sociology* 61(4), 427–445.

Gross, N. (1946) Sociological variables and cultural configurations in contemporary rural communities. PhD dissertation, Iowa State University, Ames, Iowa.

Gross, N. (1948a) Sociological variation in contemporary rural life. *Rural Sociology* 13(3), 256–273.

Gross, N. (1948b) Cultural variables in rural communities. *American Journal of Sociology* 53(5), 344–350.

Hanson, M., Goudy, W.J., Miller, R. and Whetstone, S. (1999) *Agriculture in Iowa: Trends from 1935 to 1997.* Census Services, Iowa State University, Ames, Iowa.

Hassinger, E.W. and Pinkerton, J.R. (1986) *The Human Community.* Macmillan Publishing, New York.

Hawley, A. (1950) *Human Ecology: a Theory of Community Structure.* The Ronald Press Company, New York.

Hay, D.G., Ensminger, D., Miller, S.R. and Lebrum, E.J. (1949) *Rural Organizations in Three Towns.* Bulletin 390, AES, University of Maine, Orono, Maine.

Hibbard, M. and Davis, L. (1986) When

the going gets tough: Economic reality and the cultural myths of small-town America. *Journal of the American Planning Association* (Autumn), 419–428.

Hillery, G.A., Jr (1955) Definitions of community: areas of agreement. *Rural Sociology* 20(2), 111–123.

Hillery, G.A. (1968) *Communal Organizations: a Study of Local Societies.* University of Chicago Press, Chicago, Illinois.

Hollingshead, A.B. (1950) Class differences in family stability. *Annals of the American Academy of Political and Social Science* 272, 39–46.

Hostetler, J.A. (1980) *Amish Society*, 3rd edn. The Johns Hopkins University Press, Baltimore, Maryland.

Hostetler, J.A. (1990) *Amish Society*, 4th edn. The Johns Hopkins University Press, Baltimore, Maryland.

Hovinen, G.R. (1978) Lancaster's streetcar suburbs, 1890–1920. *Journal of the Lancaster County Historical Society* 82(1), 49–59.

Intelligencer Journal (1985) p.10.

Jakle, J.A. (1982) *The American Small Town: Twentieth Century Place Images.* Archon Books, Hamden, Connecticut.

Kasarda, J.D. and Janowitz, M. (1974) Community attachment in mass society. *American Sociological Review* 39(3), 328–339.

Kaufman, H.F. (1959) Toward an interactional conception of community. *Social Forces* 38(1), 8–17.

Kemmis, D. (1990) *Community and the Politics of Place.* University of Oklahoma Press, Norman, Oklahoma.

Kollmorgen, W.M. (1942) *Culture of a Contemporary Rural Community: the Old Order Amish of Lancaster County, Pennsylvania.* Rural Life Studies Series 4, USDA, Bureau of Agricultural Economics, Washington, DC.

Konig, R. (1968) *The Community.* (Translated by E. Fitzgerald.) Schocken Books, New York.

Kraybill, D.B. (1989) *The Riddle of Amish Culture.* The Johns Hopkins University Press, Baltimore, Maryland.

Kruger, L.E. and Shannon, M.A. (2000) Getting to know ourselves and our places through participation in civic social assessment. *Society and Natural Resources* 13, 461–478.

Kunstler, J.H. (1996) *Home From Nowhere: Remaking Our World for the 21st Century.* Simon and Schuster, New York.

Leonard, O.E. and Loomis, C.P. (1941) *Culture of a Contemporary Rural Community: El Cerrito, New Mexico.* Rural Life Studies 1, USDA, Bureau of Agricultural Economics, Washington, DC.

Locke, H. (1945) Contemporary American farm families. *Rural Sociology* 10(2), 142–151.

Loomis, C.P. (1941) Informal groupings in a Spanish-American village. *Sociometry* 4(1), 26–51.

Loomis, C.P. (1958) El Cerrito, New Mexico: a changing village. *New Mexico Historical Review* 33(2), 53–75.

Loomis, C.P. (1959) Systemic linkage of El Cerrito. *Rural Sociology* 24(1), 54–57.

Loomis, C. (1960) *Social Systems: Essays on their Persistence and Change.* Van Nostrand Co., Princeton, New Jersey.

Loomis, C.P. and Beegle, J.A. (1957a) *Rural Sociology: the Strategy of Change.* Prentice-Hall, Englewood Cliffs, New Jersey.

Loomis, C.P. and Beegle, J.A. (1957b) Locality systems. In: Loomis, C.P. and Beegle, J.A. (eds) *Rural Sociology: the Strategy of Change.* Prentice-Hall, Englewood Cliffs, New Jersey, pp. 22–58.

Luloff, A.E. (1990) Communities and social change: how do small communities act? In: Luloff, A.E. and Swanson, L.E. (eds) *American Rural Communities*. Westview Press, Boulder, Colorado, pp. 214–227.

Luloff, A.E. and Befort, W.A. (1989) Land use change and aerial photography: lessons for applied sociology. *Rural Sociology* 54(1), 92–105.

Luloff, A.E. and Swanson, L.E. (1995) Community agency and disaffection: enhancing collective resources. In: Beaulieu, L. and Mulkey, D. (eds) *Investing in People: the Human Capital Needs of Rural America*. Westview Press, Boulder, Colorado, pp. 351–372.

Luloff, A.E. and Wilkinson, K.P. (1990) Community action and the National Rural Development Agenda. *Sociological Practice* 8, 48–57.

Lynd, R.S. and Lynd, H.M. (1929) *Middletown*. Harcourt, Brace and Co., New York.

Lynd, R.S. and Lynd, H.M. (1937) *Middletown in Transition: a Study of Cultural Conflicts*. Harcourt, Brace and Co., New York.

MacLeish, K. and Young, K. (1942) *Culture of a Contemporary Rural Community: Landaff, New Hampshire*. Rural Life Studies 3, USDA, Bureau of Agricultural Economics, Washington, DC.

Marsden, T., Murdoch, J., Lowe, P., Munton, R. and Flynn, A. (1993) *Constructing the Countryside*. Westview Press, Boulder, Colorado.

Mays, W.E. (1968) *Sublette Revisited: Stability and Change in a Rural Kansas Community*. Florham Park Press, New York.

McGranahan, D.A. (1999) *Natural Amenities Drive Rural Population Change*. Agricultural Economic Report No. 781, Food and Rural Economics Division, Economic Research Service, USDA, Washington, DC.

Meyers, T.J. (1993) Education and Schooling. In: Kraybill, D.B. (ed.) *The Amish and the State*. Johns Hopkins University Press, Baltimore, Maryland, pp. 87–108.

Mieher, S. (1987) Life along main street. *Georgia Trends* June, 47–55.

Miner, H. (1949) *Culture and Agriculture: an Anthropological Study of a Cornbelt County*. University of Michigan Press, Ann Arbor, Michigan.

Moe, E.O. and Taylor, C.G. (1942) *Culture of a Contemporary Rural Community: Irwin, Iowa*. Rural Life Studies 5, USDA, Bureau of Agricultural Economics, Washington, DC.

Morris, D. and Hess, K. (1975) *Neighborhood Power*. Beacon Press, Boston, Massachusetts.

New Mexico Bureau of Business and Economic Research (1999) City populations in New Mexico: 1910–1998. *New Mexico Business: Current Economic Report* 20(6), 1, 8. The University of New Mexico, Albuquerque.

North, D.C. (1955) Location theory and regional economic growth. *Journal of Political Economy* 63, 240–255.

Nostrand, R.L. (1982) El Cerrito revisited. *New Mexico Historical Review* 57(2), 1009–1122.

Nostrand, R.L. (1992) *The Hispano Homeland*. University of Oklahoma Press, Norman, Oklahoma.

Olshan, M.A. (1981) Modernity, the Folk Society, and the Old Order Amish: an alternative Interpretation. *Rural Sociology* 46(2), 297–309.

Overdest, C. (2000) Participatory democracy, representative democracy, and the nature of diffuse and concentrated interests: a case study of public involvement on a national forest district. *Society and Natural Resources* 13, 685–696.

Padfield, H. (1980) The expendable rural community and the denial of powerlessness. In: Gallaher, A., Jr and Padfield, H. (eds) *The Dying Community.* University of New Mexico Press, Albuquerque, New Mexico, pp. 159–185.

Parsons, T. (1951) *The Social System.* The Free Press, Glencoe, Illinois.

Place, E. (1993) Land use. In: Kraybill, D.B. (ed.) *The Amish and the State.* The Johns Hopkins University Press, Baltimore, Maryland, pp. 191–212.

Ploch, L.A. (1951) Factors related to the persistencies and changes in the social participation of household heads, Howard Community, Pennsylvania, 1937 and 1949. Unpublished MS thesis, Pennsylvania State University, University Park, Pennsylvania.

Ploch, L.A. (1978) The reversal in inmigration trends – some rural development consequences. *Rural Sociology* 42(2), 293–303.

Ploch, L.A. (1985) Migration and social participation. In: Steahr, T. and Luloff, A.E. (eds) *The Structure and Impact of Population Redistribution in New England.* Northeast Regional Center for Rural Development, University Park, Pennsylvania, pp. 91–109.

Ploch, L.A. (1989) *Landaff – Then and Now.* Maine Agricultural Experiment Station Miscellaneous Publication 828, University of Maine, Orono, Maine.

Poplin, D.E. (1979) *Communities: a Survey of Theories and Methods of Research.* Macmillan, New York.

Raper, A. (1936) *Preface to Peasantry.* The University of North Carolina Press, Chapel Hill, North Carolina.

Redfield, R. (1941) *The Folk Culture of Yucatan.* University of Chicago Press, Chicago, Illinois.

Reisner, M. (1993) *Cadillac Desert: the American West and its Disappearing Water.* Penguin Books, New York.

Sanders, I. (1950) *The Community.* Ronald Press, New York.

Sanders, I. (1958) Theories of community development. *Rural Sociology* 23(1), 1–12.

Sanders, I. (1966). *The Community: an Introduction to a Social System.* The Ronald Press, New York.

Schmalenbach, H. (1961) The sociological category of communion. In: Parsons, T., Shils, E., Naegele, K.D. and Pitts, J.R. (eds) *Theories of Society: Foundations of Modern Sociological Theory.* The Free Press, New York, pp. 331–347.

Sherif, M. (1966) *In Common Predicament: Social Psychology of Intergroup Conflict and Cooperation.* Houghton-Mifflin, Boston, Massachusetts.

Smith, M.D., Krannich, R.S. and Hunter, L.M. (2001) Growth, decline, stability and disruption: a longitudinal analysis of social well-being in four western communities. *Rural Sociology* 66(3), 425–450.

Smith, S.M., Findeis, J.L., Kraybill, D.B., Nolt, S.M., Kanagy, C.L. and Kozimor, M.L. (1994) *Amish Micro-Enterprises: Models for Rural Development.* Final Report, Department of Agricultural Economics and Rural Sociology, The Pennsylvania State University, University Park, Pennsylvania.

State Occupational Information Coordinating Committee of NH (1997) 1997 community profile for Landaff. http://www.state.nh.us/soiccnh/landaf.htm

Stein, M. (1960) *The Eclipse of*

Community. Princeton University Press, Princeton, New Jersey.

Stinner, W.F., Van Loon, M., Chung, S.W. and Byun, Y. (1990) Community size, individual social position and community attachment. *Rural Sociology* 55(4), 494–521.

Stone, K., Artz, G., Burke, S.C., Goudy, W.J. and Hanson, M. (1999) Retail sales in Iowa counties: 1980–1998. In: Goudy, W.J., Burke, S.C. and Hanson, M. (eds) *Iowa Counties: Selected Population Trends, Vital Statistics, and Socioeconomic Data*. Iowa Census Services, Iowa State University, Ames, Iowa, pp. 301–317.

Stranahan, S.Q. (1993) *Susquehanna: River of Dreams*. The Johns Hopkins University Press, Baltimore, Maryland.

Summers, G.F. and Clemente, F. (1976) Industrial development, income distribution, and public policy. *Rural Sociology* 41(2), 248–268.

Swanson, L.E. (2001) Rural policy and direct local participation: democracy, inclusiveness, collective agency, and locality-based policy. *Rural Sociology* 66(1), 1–21.

Taylor, C.C. (1942a) Foreword. In: Bell, E. (ed.) *Culture of a Contemporary Rural Community: Sublette, Kansas*. Rural Life Studies 2, USDA, Bureau of Agricultural Economics, Washington, DC.

Taylor, C.C. (1942b) Foreword. In: MacLeish, K. and Young, K. (eds) *Culture of a Contemporary Rural Community: Landaff, New Hampshire*. Rural Life Studies 3, USDA, Bureau of Agricultural Economics, Washington, DC.

Taylor, C.C. (1945) Techniques of community study and analysis as applied to modern civilized societies. In: Linton, R. (ed.) *The Science of Man in the World Crisis*. Columbia University Press, New York, pp. 416–441.

Taylor, C.C., Loomis, C.P., Provinse, J., Huett, J.E., Jr and Young, K. (1940) *Cultural, Structural and Social-Psychological Study of Selected American Farm Communities: Field Manual*. USDA, Bureau of Agricultural Economics, Washington, DC.

Testa, R.M. (1992) *After the Fire*. University Press of New England, Hanover, New Hampshire.

The Daily Nonpareil (1992) Irwin–Kirkman, Manilla Set to Merge. January 29.

Tonnies, F. (1957) *Community and Society (Gemeinschaft und Gesselschaft)*. (Translated and edited by C.P. Loomis.) Michigan State University Press, East Lansing, Michigan.

US Census of Population and Housing (1990a) *Summary Tape File (STF) 3A*. US Government Printing Office, Washington, DC.

US Census of Population and Housing (1990b) *Summary Tape File (STF) 1A*. US Government Printing Office, Washington, DC.

United States Department of Agriculture (1999) *1997 Census of Agriculture: Iowa, State and County Data*. National Agricultural Statistics Service, Washington, DC.

Vidich, A.J. and Bensman, J. (1968) *Small Town in Mass Society: Class, Power and Religion in a Rural Community*. Princeton University Press, Princeton, New Jersey.

Vogt, E. and O'Dea, T. (1953) A comparative study of the role of values in social action in two southwestern communities. *American Sociological Review* 18 (December), 645–654.

Warner, W.L. (1963) *Yankee City*. Yale University Press, New Haven, Connecticut.

Warner, W.L., Meeker, M. and Eels, K. (1949) *Social Class in America: the*

Evaluation of Status. Harper Torchbooks, New York.

Warren, R.L. (1963) *The Community in America.* Rand McNally, Chicago, Illinois.

Warren, R.L. (1970) The good community – what would it be? *Journal of the Community Development Society* 1(1), 14–23.

Warren, R.L. (1972) *The Community in America,* 2nd edn. Rand McNally, Chicago, Illinois.

Warren, R.L. (1978) *The Community in America,* 3rd edn. Rand McNally, Chicago, Illinois.

Weber, E.P. (2000) A new vanguard for the environment: grass-roots ecosystem management as a new environmental movement. *Society and Natural Resources* 13, 237–259.

Wellman, B. (1979) The community question: the intimate networks of East Yorkers. *American Journal of Sociology* 84(5), 1201–1231.

Wellman, B. and Wortley, S. (1990) Different strokes from different folks: community ties and social support. *American Journal of Sociology* 96(3), 558–588.

West, J. (1945) *Plainville, U.S.A.* Columbia University Press, New York.

Wilkinson, K.P. (1970a) Phases and roles in community action. *Rural Sociology* 35(1), 54–68.

Wilkinson, K.P. (1970b) The community as a social field. *Social Forces* 48(3), 311–322.

Wilkinson, K.P. (1979) Social well-being and community. *Journal of the Community Development Society* 10(1), 5–16.

Wilkinson, K.P. (1986) In search of the community in the changing countryside. *Rural Sociology* 51(1), 1–17.

Wilkinson, K.P. (1990) Crime and community. In: Luloff, A.E. and Swanson, L.E. (eds) *American Rural Communities.* Westview Press, Boulder, Colorado, pp. 151–168.

Wilkinson, K.P. (1991) *The Community in Rural America.* Greenwood Press, Westport, Connecticut.

Williams, D.D. (1995) From dust bowl to green circles: a case study of Haskell County, Kansas. Unpublished PhD dissertation, Kansas State University, Manhattan, Kansas.

Wilson, W.J. (1987) *The Truly Disadvantaged: the Inner City, the Underclass and Public Policy.* University of Chicago Press, Chicago, Illinois.

Wirth, L. (1938) Urbanism as a way of life. *American Journal of Sociology* 44(1), 1–24.

Wynne, W. (1943) *Culture of a Contemporary Rural Community: Harmony, Georgia.* Rural Life Studies 6, Bureau of Agricultural Economics, Washington, DC.

Index

agricultural and natural resource dependency 6
American Sociological Association 2
Anderson, Sherwood 11
   Winesburg, Ohio 11

Bauer, R.A. and Bauer, A.H. 7
Bell, Earl 7, 23

City and Community 2
community and family stability 6
community theory 9
community-based forestry 2
comparative analysis 5
Corn Belt community 45
   see also Irwin, Iowa
cultural isolation 5
Culture of a Contemporary Rural Community, The 3

Demise of community 1
Dewey, Richard 14
Durkheim, Emile 10

El Cerrito (New Mexico) 3, 4, 5, 117–141, 172, 173, 174, 175, 176
   acequia 120, 123, 125, 126, 131, 133, 174
   agriculture 119, 122
   community actions 130–133
   community background 117–119
   community and social well-being 133–138
   family 125
   funcion 123
   leadership 124–125
   limpia 123–124, 125, 126, 140, 173
   local society 129–130
   locality 126–128
   population 120–126
   San Miguel County 117
   social institutions 120, 123

family systems 5
First World War 11
Ford, Henry 11

Gallaher, A. 6
Great Change thesis *see* Warren, Roland
Great Depression 3, 171
Gross, N. 5

Harmony (Georgia) 3, 4, 5, 71–93, 172, 173, 174, 175, 176
  agriculture 71–72
  community institutions 75–77
  community life 73–75, 84–87
  economic structure 79–84
  lake residents 89–91
  Putnam County 71–93
    Lake Oconee dam project 74, 174
  race relations 76–77, 87–88
Hollingshead, A.B. 6
  Elmstown (North Carolina) 6
Howard (Pennsylvania) 6

industrialization 6
interactional perspective 41–42
intracommunity interaction systems 5
Irwin (Iowa) 3, 4, 5, 45–70, 172, 173, 174, 175, 176
  agriculture 49–52
    agriculture/community relationship 52–55
  community action 66–68
  Corn Belt community 45
  institutional structure 56–62
  physical characteristics 47–48
  Shelby County 45–70
  social system 62–66

Jonesville (North Carolina) 6

Kemmis, D. 2
Kollmorgen, Walter 5, 143

Landaff (New Hampshire) 3, 4, 5, 95–116, 172, 173, 174, 175, 176
  Blue School 111–112, 173
  community characteristics 98–100
  community resource base 102–105
  government 112–113
  Grafton County 98
  population 98, 100–102, 107–109
  social organizations 105–107, 109–112

land use and environmental issues 6
Llewellyn Park (New Jersey) 14
locality-based civic engagement 2
Locke, H. 4
Loomis, C.P. 5
Luloff, A.E. 7
Lynd, R.S. and Lynd, H.M. 1, 6, 7
  Middletown 6, 8

mass society thesis 7
Mays, W.E. 5, 23
Medill, Joseph 15

'New Deal' programmes 27
Nostrand, R.L. 5

Old Order Amish (Lancaster County, Pennsylvania) 3, 4, 5, 143–170, 172, 173, 174, 175, 176
  agriculture 143, 146–149, 164
  autonomy 147–155
    government intrusion 152
    local regulations 151
    schooling 154–155
    self-employment 151
    social security 150, 151
  demographic changes 145–146
  Lancaster County 3, 4, 5, 143–170, 174, 175, 176
  land use changes 145–146, 153
  Mennonites 163
  population 145–146, 167–168
  psychological identification 162–165
  sense of community 157–159
  service areas 159–161
  tourism 166
Olshan, M.A. 5
Overdest, C. 2

Parsons, Talcott 9
Plainville (Massachusetts) 5, 6
Ploch, Louis A. 4, 5, 6
Poplin, D.E. 5

Redfield, R. 1
religious systems 5
rural family 4
rural industrialization 6
Rural Life Study series 3, 6, 7, 8, 10, 23, 45, 71, 95, 117, 143, 171
    El Cerrito (New Mexico) 3, 4, 5, 117–141, 172, 173, 174, 175, 176
    Harmony (Georgia) 3, 4, 5, 71–93, 172, 173, 174, 175, 176
    Irwin (Iowa) 3, 4, 5, 45–70, 172, 173, 174, 175, 176
    Landaff (New Hampshire) 3, 4, 5, 95–116, 172, 173, 174, 175, 176
    Old Order Amish (Lancaster County, Pennsylvania) 3, 4, 5, 143–170, 172, 173, 174, 175, 176
    Sublette (Kansas) 3, 4, 5, 23–43, 172, 173, 174, 175, 176
Rural Sociological Society 2
Rural Sociology 2
rural–urban dichotomy 5

Sanders, I. 5
Schmalenbach, Herman 17
    bund 17
Second World War 10, 11, 14
Society and Natural Resources 2
Stein, M. 1
Sublette (Kansas) 3, 4, 5, 23–43, 172, 173, 174, 175, 176
    agriculture 24, 25, 29–31
    community views of government 27–29
    economic development 25
    feedlots and packing plants 25
    Haskell County 23–43
    human psychology and social ties 31–33
    irrigation 24, 25
    local autonomy 33
    organization and change 33–39
    water issues 25
    weather 24
suburbs/suburbanization 14
Swanson, Louis 2

Taylor, Carl 3, 4
Tonnies, F. 1, 10
    gemeinschaft 10, 16, 18
    gessellschaft 10, 18
    natural will vs. rational will 17
Twain, Mark 1

United States Department of Agriculture 3, 4, 45, 71, 95

Warner, W.L. 6, 7, 8
Warren, Roland 4, 9, 10, 11
    Community in America, The 17
    Great Change thesis 10, 11, 12, 13, 14, 15, 16, 17, 39–40, 173
Weber, E.P. 2
Wellman, B. 16
Wellman, B. and Wortley, S. 16
West, James 5
Wilkinson, Kenneth 1, 8, 11, 19, 143, 174, 175, 176
Wirth, Louis 1
    essay on urbanism 14